Adiabatic Quantum Computation and Quantum Annealing

Theory and Practice

Synthesis Lectures on Quantum Computing

Editor
Marco Lanzagorta, *U.S. Naval Research Labs*
Jeffrey Uhlmann, *University of Missouri-Columbia*

Adiabatic Quantum Computation and Quantum Annealing: Theory and Practice

Catherine C. McGeoch

ISBN: 978-3-031-01390-4   paperback
ISBN: 978-3-031-02518-1   ebook

DOI 10.1007/978-3-031-02518-1

A Publication in the Springer series
SYNTHESIS LECTURES ON QUANTUM COMPUTING

Lecture #8
Series Editors: Marco Lanzagorta, *U.S. Naval Research Labs*
                         Jeffrey Uhlmann, *University of Missouri–Columbia*
Series ISSN
Print 1945-9726   Electronic 1945-9734

Adiabatic Quantum Computation and Quantum Annealing

Theory and Practice

Catherine C. McGeoch
Amherst College
D-Wave Systems Inc.

SYNTHESIS LECTURES ON QUANTUM COMPUTING #8

ABSTRACT

Adiabatic quantum computation (AQC) is an alternative to the better-known gate model of quantum computation. The two models are polynomially equivalent, but otherwise quite dissimilar: one property that distinguishes AQC from the gate model is its analog nature. Quantum annealing (QA) describes a type of heuristic search algorithm that can be implemented to run in the "native instruction set" of an AQC platform.

D-Wave Systems Inc. manufactures quantum annealing processor chips that exploit quantum properties to realize QA computations in hardware. The chips form the centerpiece of a novel computing platform designed to solve NP-hard optimization problems. Starting with a 16-qubit prototype announced in 2007, the company has launched and sold increasingly larger models: the 128-qubit D-Wave One system was announced in 2010 and the 512-qubit D-Wave Two system arrived on the scene in 2013. A 1,000-qubit model is expected to be available in 2014.

This monograph presents an introductory overview of this unusual and rapidly developing approach to computation. We start with a survey of basic principles of quantum computation and what is known about the AQC model and the QA algorithm paradigm. Next we review the D-Wave technology stack and discuss some challenges to building and using quantum computing systems at a commercial scale. The last chapter reviews some experimental efforts to understand the properties and capabilities of these unusual platforms. The discussion throughout is aimed at an audience of computer scientists with little background in quantum computation or in physics.

KEYWORDS

quantum computing, adiabatic quantum computation, quantum annealing, D-Wave Systems, optimization, NP-hard problems

Contents

Acknowledgments

I would like to express my very great appreciation to those who patiently answered my many questions, made thoughtful suggestions about improving draft manuscripts, and contributed just the right images for conveying complex ideas: Mohammad Amin, Andrew Appel, Paul Bunyk, Carrie Cheung, Mark Johnson, Jeremy Hilton, Andrew King, Jamie King, Trevor Lanting, Bill Macready, Lyle McGeoch, Miles Steininger, John Rager, Geordie Rose, Aidan Roy, and Murray Thom.

Catherine C. McGeoch
July 2014

CHAPTER 1

Introduction

In May 1919, Professor David Todd of Amherst College, an astronomer and foremost authority on eclipses, made national headlines by announcing a bold plan for his next observation expedition to the coast of Uraguay: to photograph the upcoming solar eclipse using a camera mounted on an airplane! Besides eliminating any chance that the expedition would fail due to cloud cover, he posited that photos taken from high in the atmosphere would yield the clearest by far solar corona images ever seen.

Expert reactions were mixed as to whether this plan could possibly succeed. Professor Harold Jacoby of Columbia University, while generally optimistic, pointed out several obstacles. One that seemed especially insurmountable was how to fit a camera of sufficient fidelity onto an airplane: "A camera with a fourteen-foot focus takes a picture which shows the sun only as large as a quarter, and I doubt the possibility of manipulating a machine of even this size."[1] As it turned out the experiment failed for an unexpected reason: the plane was destroyed in a hurricane on the coast of Brazil just before the eclipse date [79].

Cameras have gotten a lot smaller and airplanes a lot bigger since then, and Prof. Jacoby's obstacle is no longer considered insurmountable. But his remarks illustrate a predicament that arises whenever science confronts new technology—whenever theory meets practice: it can be very difficult to tell, in the moment, which obstacles to success are fundamental and insurmountable, and which are merely transient engineering hurdles.

That predicament returns in this new century, when—prompted by announcements by D-Wave Systems Inc. of the production and sale of working computing systems with chips containing up to 512 quantum bits (qubits)—scientists are asked to make predictions about the potential capabilities of this new technology.

Plausibility arguments from complexity theory tell us that quantum computers may be able to achieve enormous speedups over classical models of computation on some problems. But this question is as open as P vs. NP: experts may reasonably disagree as to the eventual outcome. Experts may also point out that attempts in physics laboratories worldwide to build working quantum computers of interesting size have so far revealed very significant obstacles to success. But much of the accepted knowledge about what is necessary for quantum computation to succeed in practice has been focused on the goal of achieving universal computation in the gate model, and

---

[1]Prof. Todd conceded that there would probably be no room for both the camera and a passenger (himself), but he had worked out a scheme whereby the pilot could aim and operate the camera while flying the biplane-seaplane with engines turned off to eliminate vibrations.

does not necessarily apply to these unusual platforms. D-Wave chip are designed along the lines of the *adiabatic* model of quantum computation (AQC), an alternative to the better-known quantum gate model. They are not full CPUs,[2] but rather more like "smart memory accelerators," designed to solve NP-hard optimization problems using a solution method called *quantum annealing* (QA). When supercooled to temperatures below 20mK (very near absolute zero), the chips are able to exploit quantum properties to carry out their computations.

Theoretical understanding of computation in the AQC framework has developed steadily since its proposal by Farhi et al. [36] (see also Farhi et al. [35]). Research on the QA paradigm, originally proposed for implementation on classical platforms,[3] has also proceeded apace. And since around 2007, experimentalists have been able to study this approach to problem-solving on platforms that are purpose-built. A good-sized body of experimental work has now been published that examines succeeding models of these systems, and a rough outline of their properties has begun to emerge. Nevertheless many questions about the potential power of this new approach to computation remain wide open.

This is perhaps not surprising given the timeline of developments. D-Wave chip sizes have grown from a 16-qubit prototype announced in 2007, to a 128-qubit chip in 2010, a 512-qubit chip in 2013, and a 1,152-qubit chip projected for release in late 2014. Larger new models are typically faster and better (lower error rates) than previous ones. Thus experimental work on any given chip tends to have a very short shelf-life.

By comparison, consider that one of Intel's first products, released in 1969, was an 8-byte SRAM chip; in 1971 they announced a 40-byte RAM chip; and in 1976 they launched the 8,048 "computer on a chip" with 64 bytes onboard and 256 bytes of external RAM. There was rapid development in both cases; and certainly no amount of experimentation on those early Intel products could have foretold the advances that lay ahead.

This monograph presents an introduction to the theoretical underpinnings, practical design challenges, and emerging performance profiles surrounding this novel approach to computation. Space limitations prohibit full development of these ideas starting from first principles; nor can this be an exhaustive survey of the rapidly growing body of literature in the field. Rather the focus is on introducing the central ideas and summarizing the major research highlights, while providing many references to further reading.

The discussion is aimed at readers with graduate-level training in computer science. We draw on material from a variety of areas including complexity theory, algorithms, computer architecture, concurrent systems, combinatorial optimization, and experimental algorithmics. A rudimentary familiarity with linear algebra and elementary concepts of physics is also assumed. No background in quantum mechanics or quantum computing is required.

---

[2]To clarify a point that is often misunderstood: While the chips alone are not universal computers, the product the company sells—and sells time on via the cloud—is an integrated computing system that includes a conventional processor, a quantum annealing chip and peripherals housed in a cyrogenic refrigeration chamber, and a software bundle with components ranging from low-level controllers to program development tools.

[3]The idea appears to have been developed independently by researchers in several disciplines.

1.1   WHAT'S INSIDE

We start with a quick introduction to the terminology and key ideas that will be developed in later chapters.

Quantum computation.   Interest in the idea of building quantum computers was sparked by Feynman in a talk about using them to simulate quantum mechanical processes (Feynman [37]). Soon after, Deutsch [28] developed a universal model of computation—the *quantum gate model*—that can simulate any Turing machine computation with no more than polynomial cost overhead. The interesting open question is whether the reverse is true: can a Turing machine efficiently simulate a quantum computer?

Quantum computers operate on *qubits*, which exhibit peculiar properties such as *superposition* and *entanglement*. Optimism about a so-called *quantum speedup* is based on the idea that quantum algorithms might be able to exploit these properties to carry out a computation in asymptotically fewer basic operations than is possible (as defined by a lower bound on problem complexity) for any classical algorithm. For example, Grover's algorithm [45] can search an unordered database of size n in $O(\sqrt{n})$ time in a quantum model of computation, whereas every classical algorithm needs $\Omega(n)$ time to at least visit every element in the database. This represents a quantum speedup on the order of root-n.

Another example of a problem that may yield quantum speedup is factoring. Shor's algorithm (Shor [99]) can factor an n-bit number in polynomial time on a quantum computer in the gate model. The lower bound for factoring in the classical model is not known, but the best-known classical algorithms take super-polynomial time. Since the security of RSA encryption (as currently implemented) is based on the presumed intractability of factoring, Shor's result—sometimes described as the killer app for quantum computing—has sparked world-wide efforts to develop working quantum computers.

For comprehensive introductions to quantum computation in the gate model, see Nielsen and Chuang [75] or Rieffel and Polak [88]. Chapter 2 introduces basic concepts of quantum computing and gives an overview of relevant complexity classes.

AQC.   *Adiabatic quantum computation*[4] (AQC) is an alternative to the quantum gate model. Results by Farhi et al. [35] and Aharonov et al. [1] imply that the two models are polynomially equivalent. Note that quantum computation is probabilistic, so that instead of analyzing the computation time of a given algorithm, we analyze the tradeoff between time and the probability that the output is correct.

One distinction between the two models is their discrete versus analog natures, which creates a very different look-and-feel for algorithms, as illustrated in Figure 1.1. Panel (a) shows a simplified diagram of the Deutsch-Josza algorithm, one of the first efficient algorithms created

---

[4]*Adiabatic* refers to a thermal (not quantum) process that experiences no transfer of energy between the system and its environment. The term is used metaphorically in this context to describe a quantum computation that does not interact with its environment.

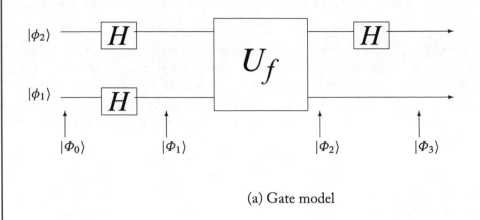

$|\phi_2\rangle$ H U_f H

$|\phi_1\rangle$ H

$|\Phi_0\rangle$ $|\Phi_1\rangle$ $|\Phi_2\rangle$ $|\Phi_3\rangle$

(a) Gate model

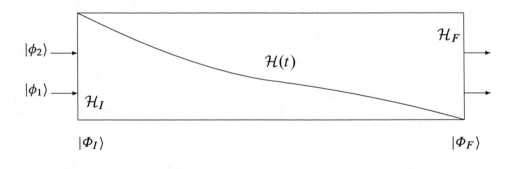

$|\phi_2\rangle$

$|\phi_1\rangle$ $\mathcal{H}_I$ $\mathcal{H}(t)$ $\mathcal{H}_F$

$|\Phi_I\rangle$ $|\Phi_F\rangle$

(b) AQC model

Figure 1.1: (a) Diagram of an algorithm in the gate model. (b) Diagram of an algorithm in the AQC model.

for the gate model (see Rieffel and Polak [88] or Deutsch and Jozsa [29] for more). It resembles a classical circuit diagram: the horizontal lines follow two qubits through time, and the boxes represent quantum circuits that change qubit states. The notations along the bottom mark computation states at discrete times in the process, before and after circuits are applied.

Panel (b) depicts an analog AQC computation. There are no gates nor discrete time steps: instead the qubit states gradually evolve according to certain forces represented by *Hamiltonians*. The algorithm performs a transition from the initial Hamiltonian $\mathcal{H}_I$ to the final Hamiltonian

$\mathcal{H}_F$, and the qubit states are read at the end of the transition. Each observable n-qubit state has an associated *energy*; the *ground state* has the lowest energy over all states. According to the Adiabatic Theorem (Born and Fock [16]), if the qubits start in ground state and the transition is carried out slowly enough, then with high probability the system will finish in ground state. The idea is to define $\mathcal{H}_I$ so that a ground state is easy to find, and to define $\mathcal{H}_F$ so that its ground state corresponds to an optimal solution to a given optimization problem P.

These concepts are developed further in Chapter 2, which presents some simple AQC algorithms and their analyses together with an overview of relevant complexity classes.

Quantum annealing.   The term *quantum annealing* describes a heuristic search approach to solving problems in combinatorial optimization. Quantum annealing bears some resemblance to the better-known simulated annealing (SA) heuristic, in that both are intended to mimic natural physical processes.

Quantum annealing algorithms have the handy property of being expressible in the "native instruction sets" of both classical and AQC models of computation. Many authors now use the terms QA and AQC interchangeably when discussing these algorithms, although in this text we observe a technical distinction described in Chapter 4.

This connection between optimization and quantum computing has sparked the proposal of QA (and AQC) algorithms for several NP-hard problems as well as for other application areas (such as simulation of quantum phenomena and solving continuous optimization problems). Many of these algorithms have been implemented on classical computers; a few have been implemented to run on quantum annealing platforms built by D-Wave and by other research groups. Chapter 3 describes the quantum annealing paradigm and surveys some algorithms developed for combinatorial optimization and decision problems.

Building a working quantum annealer.   Van Meter and Horsman [107] survey the huge variety of approaches taken in physics laboratories worldwide to construct working quantum computers. They write:

> The heroic efforts of experimentalists have brought us to the point where approximately 10 qubits can be controlled and entangled. Getting to that stage has been a monumental task as the fragile quantum system must be isolated from the environment, and its state protected from drifting.

In this context, D-Wave's announcements of working quantum chips with 128 and 512 qubits have been greeted with astonishment and downright scepticism.

The explanation for this apparent disconnect has two parts. First, all of the heroic laboratory work mentioned above has been aimed at building universal computers along the lines of the quantum gate model. The adiabatic model has certain intrinsic features that make it robust against some widely recognized obstacles to success. For example, one important challenge to realizing a quantum computer is *decoherence*: qubits have a tendency to lose information due to interactions with the environment, which can decrease the success probability enough to render the computation useless. Elaborate schemes for counteracting decoherence, for example by introducing extra qubits to perform error correction, are necessary to ensure adequate success probabilities in the gate model. AQC is robust against decoherence because the computation takes place in ground state. In layman's terms, qubits naturally want to decohere "toward" lower energy states; since a ground state (the lowest energy state) is the desired outcome, chances of failure for this reason are small. Note that environmental interference does create other types of challenges, as outlined in Chapter 4.

The second part of the explanation is: design simplification from relaxing computational goals. These chips are built to solve one NP-hard optimization problem (the Ising Model) using a specific solution approach (quantum annealing). The engineering challenges to realizing successful quantum computations in this framework are significant and daunting, but also much less imposing than those faced in the above-mentioned attempts to construct universal programmable quantum computers.

Chapter 4 presents an overview of the D-Wave technology stack and surveys some practical challenges to realizing quantum annealing systems that are capable of successful computations involving hundreds (soon thousands) of qubits.

Experience with quantum computation.   D-Wave does not mass-produce its quantum platforms; only a very small number have been in operation since the first chip came online in 2004 at company headquarters in Burnaby, BC. Since 2011, a D-Wave One system (128 qubits) and later a D-Wave Two system (512 qubits) have been in operation at the USC-Lockheed Martin Quantum Computation Center (QCC). In 2013, another D-Wave Two system was installed at the Quantum Artificial Intelligence Laboratory (QuAIL), jointly sponsored by Google and NASA. Chapter 5 surveys experimental research to understand the properties of these systems. This work has been carried out by scientists and engineers at the above-named organizations, as well as by guest researchers at the three sites. Much of this experimental work has focused on three questions:

1. *What can it do?* In theory, any problem in NP can be transformed to the problem solved by the chip with at most polynomial overhead, and theoretical bounds on performance are known in some cases. But practical questions of feasible problem sizes and real success probabilities must be addressed empirically. Section 5.1 describes experience using D-Wave systems to solve problems in a variety of application domains.

2. *Is it quantum?* The question is simple enough, but experiments to demonstrate "quantumness" are difficult to devise. Section 5.2 reviews research efforts aimed at teasing out answers.

3. *How fast is it?* Referring to "it"—the chip—in this generic sense is somewhat misleading. D-Wave has committed to a rapid development process that includes realizing a new design every few months and doubling qubit counts each year: this strategy exploits the huge

room for improvement that comes from working with an immature technology. As a result, empirical performance measurements are necessarily narrow in scope and soon obsolete. Section 5.3 surveys some attempts to address this question.

1.2   WHAT HAPPENS NEXT?

These are very early days in the quantum computing era, and we are a long way from fully understanding what may be realized. But that doesn't mean one can't have opinions about future developments. See Grossman [44] or Hsu [52] for reviews of ongoing debates about how D-Wave's approach to quantum computing will play out. In a nutshell: pessimists point out some obstacles to success that appear to be insurmountable; optimists point out that many apparent obstacles have turned out to be mere engineering hurdles, and the outlook is bright. The goal in subsequent chapters is to equip the reader to make his or her own predictions. Or, as Alan Kay put it:

> *The best way to predict the future is to invent it. Really smart people with reasonable funding can do just about anything that doesn't violate too many of Newton's Laws.* (Kay [57])

Granted, the scientists and engineers in this emerging field are trying hard to violate a few of Newton's Laws, but the principle still holds. Let's find out.

CHAPTER 2

Adiabatic Quantum Computation

Adiabatic quantum computation (AQC) was introduced by Farhi et al. [36] (see also Farhi et al. [35]) as an alternative to the more familiar gate model of quantum computation. Besides the intrinsic interest surrounding any new ideas about computation, AQC shows promise as being more "realizable" and possibly more "analyzable" than the gate model via a fairly well-developed set of analysis techniques from quantum mechanics.

This chapter gives an introductory overview of this developing area. We start in Section 2.4 with basic definitions and notations of quantum computation. The gate model is only briefly discussed: for an introduction to that more established field see Nielsen and Chuang [75], Mermin [71], Rieffel and Polak [88], or Williams [108]. The emphasis here is on mathematical formalisms, with only rare attempts to justify the notational framework by explaining the mind-bending properties of quantum phenomena. See Greenstein and Zajonc [43] for an introduction to the major experiments and results of quantum mechanics.

Section 2.2 describes the general structure of AQC algorithms, with a simple example and a discussion of the Adiabatic Theorem, which drives the analysis of computation time. Section 2.3 presents a richer example together with some partial analyses. Section 2.4 gives an overview of relevant complexity classes and shows how AQC fits within that landscape.

2.1 BASICS OF QUANTUM COMPUTATION

This section introduces the mathematical framework of quantum computation and points out the key differences between quantum and classical models of computation such as Turing machines and RAMs.

Qubits and their properties. The most obvious difference between classical and quantum computation is the concept of state. The state of a classical algorithm can be represented by a register R of bits, each having value 0 or 1. The state of a quantum algorithm is represented by a register Q containing *qubits*, which have strange properties.

One such property is *superposition*, which means a qubit can be in states 0 and 1 simultaneously. A superposition state $|\phi\rangle$ (pronounced "ket phi" in the Dirac bra-ket notation) corresponds to a qubit state vector containing two complex numbers α and β. The state vector denotes a linear combination of two *basis states* chosen conventionally to be $|0\rangle$ and $|1\rangle$: that is, $|\phi\rangle = \alpha |0\rangle + \beta |1\rangle$.

In matrix notation we have

$$|\phi\rangle = \begin{pmatrix} \alpha \\ \beta \end{pmatrix}$$

$$|0\rangle = \begin{pmatrix} 1 \\ 0 \end{pmatrix}$$

$$|1\rangle = \begin{pmatrix} 0 \\ 1 \end{pmatrix}. \tag{2.1}$$

A key fact of quantum mechanics is that superposition states cannot be directly observed. Instead, when a qubit is measured with respect to a given basis, we imagine that it instantly "collapses" to one of the basis states: to $|0\rangle$ (observed as classical state 0) with probability $|\alpha|^2$, and to $|1\rangle$ (observed as 1) with probability $|\beta|^2$. Thus we have

$$|\alpha|^2 + |\beta|^2 = 1, \tag{2.2}$$

where $|\alpha|$ is the magnitude of the complex number: $|\alpha| = |x + iy| = \sqrt{x^2 + y^2}$. Basis states are also called the *observable states* of the qubit.

Another way to think about this difficult-to-understand phenomenon is to say that any measurement device defines a pair of orthogonal bases (such as $|0\rangle$, $|1\rangle$, but others may be used) on which the qubit superposition state must be projected to be read. The projection changes α and β so that it is impossible to read the same qubit state twice.

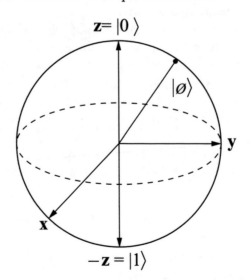

Figure 2.1: A Bloch sphere.

A state vector can be visualized as a point on the surface of the three-dimensional unit *Bloch sphere* shown in Fig 2.1. By convention the north and south poles correspond to the standard basis states $|0\rangle$ and $|1\rangle$, respectively. We will also be interested in an alternative pair of basis states $|x_+\rangle$ and $|x_-\rangle$ that lie on the positive and negative x axes.

Quantum gates and operators.   A *quantum logic gate* operating on a qubit causes a transformation that rotates the state vector to another location on the Bloch sphere. The three fundamental rotation operators are expressed by the *Pauli matrices* $\sigma^x, \sigma^y, \sigma^z$. Together with the identity transformation I these are represented in matrix notation by

$$I = \begin{pmatrix} 1 & 0 \\ 0 & 1 \end{pmatrix}$$

$$\sigma^x = \begin{pmatrix} 0 & 1 \\ 1 & 0 \end{pmatrix}$$

$$\sigma^y = \begin{pmatrix} 0 & -i \\ i & 0 \end{pmatrix}$$

$$\sigma^z = \begin{pmatrix} 1 & 0 \\ 0 & -1 \end{pmatrix}. \tag{2.3}$$

For example, it is easy to verify that applying the Pauli-x gate σ^x to a given qubit vector $|\phi\rangle = (\alpha, \beta)^T$ causes a reflection about the x, y plane so that α, β switch places. Thus we have $\sigma^x |1\rangle \rightarrow |0\rangle$ and $\sigma^x |0\rangle \rightarrow |1\rangle$. This is the quantum equivalent of a NOT gate.

One of the differences between classical and quantum gates is that quantum logic must be *reversible*. That is, for any quantum logic operator it must be possible to determine the input bits from reading the output bits. Excepting the NOT gate, this is not true of classical gates: for example knowing that the output of an OR gate is 1 is not enough to tell what the inputs are.

Eigenvectors and eigenvalues.   Continuing with our matrix representation: an *eigenvector* of a square matrix A is a non-zero vector v such that multiplying A by v yields a constant multiple of v: that is, $Av = \lambda v$ for some scalar λ. The coefficient λ is the *eigenvalue* of A corresponding to v.

The Pauli matrices each have two eigenvalues $+1, -1$. Later in this chapter we will make use of the eigenvectors and eigenvalues of σ^z:

$$\begin{aligned} \sigma^z |0\rangle &= +1 |0\rangle \\ \sigma^z |1\rangle &= -1 |1\rangle \end{aligned} \tag{2.4}$$

These equations can be verified using matrix notation as follows.

$$\begin{pmatrix} 1 & 0 \\ 0 & -1 \end{pmatrix} \begin{pmatrix} 1 \\ 0 \end{pmatrix} = +1 \cdot \begin{pmatrix} 1 \\ 0 \end{pmatrix}$$

$$\begin{pmatrix} 1 & 0 \\ 0 & -1 \end{pmatrix} \begin{pmatrix} 0 \\ 1 \end{pmatrix} = -1 \cdot \begin{pmatrix} 0 \\ 1 \end{pmatrix} \tag{2.5}$$

The Pauli-x matrix σ^x has the same eigenvalues but a different pair of eigenstates. Adopting the notation

$$|x_+\rangle = \begin{pmatrix} \frac{1}{\sqrt{2}} \\ \frac{1}{\sqrt{2}} \end{pmatrix}$$

$$|x_-\rangle = \begin{pmatrix} \frac{1}{\sqrt{2}} \\ -\frac{1}{\sqrt{2}} \end{pmatrix}, \qquad (2.6)$$

it is also straightforward to verify that

$$\begin{aligned} \sigma^x |x_+\rangle &= +1\,|x_+\rangle \\ \sigma^x |x_-\rangle &= -1\,|x_-\rangle \end{aligned} \qquad (2.7)$$

Note that $|x_+\rangle$ and $|x_-\rangle$ form an alternative pair of basis states to $|0\rangle$ and $|1\rangle$.

Collections of qubits.   Two qubit states $|\phi_1\rangle = (\alpha_1, \beta_1)^T$ and $|\phi_2\rangle = (\alpha_2, \beta_2)^T$ can be combined to form $|\Phi\rangle = |\phi_1\phi_2\rangle$ by applying the *tensor product* $\otimes$ to their state vectors:

$$(\alpha_1, \beta_1)^T \otimes (\alpha_2\beta_2)^T = (\alpha_1\alpha_2, \ \alpha_1\beta_2, \ \beta_1\alpha_2, \ \beta_1\beta_2)^T. \qquad (2.8)$$

A two-qubit state space has four basis states, represented as follows.

$$|00\rangle = \begin{pmatrix} 1 \\ 0 \end{pmatrix} \otimes \begin{pmatrix} 1 \\ 0 \end{pmatrix} = \begin{pmatrix} 1 \\ 0 \\ 0 \\ 0 \end{pmatrix}$$

$$|01\rangle = \begin{pmatrix} 0 \\ 1 \\ 0 \\ 0 \end{pmatrix}$$

$$|10\rangle = \begin{pmatrix} 0 \\ 0 \\ 1 \\ 0 \end{pmatrix}$$

$$|11\rangle = \begin{pmatrix} 0 \\ 0 \\ 0 \\ 1 \end{pmatrix}. \qquad (2.9)$$

Setting $\gamma_1 = \alpha_1\alpha_2$, $\gamma_2 = \alpha_1\beta_2$, $\gamma_3 = \beta_1\alpha_2$ and $\gamma_4 = \beta_1\beta_2$, we can represent $|\Phi\rangle$ in terms of its bases as follows.

$$|\Phi\rangle = |\phi_1\phi_2\rangle \;=\; (\gamma_1, \gamma_2, \gamma_3, \gamma_4)^T$$

$$=\; \gamma_1\,|00\rangle + \gamma_2\,|01\rangle + \gamma_3\,|10\rangle + \gamma_4\,|11\rangle, \quad \text{where}$$

$$1 \;=\; |\gamma_1|^2 + |\gamma_2|^2 + |\gamma_3|^2 + |\gamma_4|^2. \tag{2.10}$$

These values represent the probabilities that the qubit pair will be observed in each possible state.

Entanglement.   Another key property that distinguishes quantum from classical computation is *entanglement*: quantum particles such as qubits can interact in such a way that after the interaction, they are no longer independent. That is, their α's and β's are linked so that the probability of observing, say, $|\phi_1 = 1\rangle$ is correlated with the probability of observing $|\phi_2 = 1\rangle$.

A famous example of this phenomenon is the *Bell state*, which describes an entanglement according to

$$\frac{1}{\sqrt{2}} \left(|00\rangle + |11\rangle \right). \tag{2.11}$$

The probability of observing $|01\rangle$ or $|10\rangle$ is zero, and the probabilities of observing $|00\rangle$ or $|11\rangle$ are each 1/2. Note that each individual qubit has an equal probability of being read as $|0\rangle$ or $|1\rangle$. Because of entanglement, probabilities associated with multi-qubit states like $|00\rangle$ cannot in general be decomposed into products of individual probabilities.

Another weird property of quantum mechanics is that entanglement between quantum particles can persist even when the particles are physically separated. Entanglement also persists in time, through gate transformations and measurement. So, for example, even though entangled qubit ϕ_1 is measured first and its value destroyed, the correlation can persist until ϕ_2 is measured. There is no analog to entanglement in classical computation.

Quantum circuits.   Suppose we have a quantum register Q containing n qubits. The superposition state of Q is represented by a vector $|\Phi\rangle$ of length $N = 2^n$.

Deutsch [28] showed that a quantum circuit can be built of reversible quantum logic gates to compute any classical function f defined on n bits. Let C_f denote such a circuit applied to register Q: the result is a new superposition state, $C_f\,|\Phi\rangle \to |\Phi'\rangle$.

A given circuit C_f can be represented as an $N \times N$ matrix. For example the *controlled NOT gate* (C_{cnot}) takes two qubits as inputs: if the first (control) qubit is 1 it toggles the second (target) qubit, and otherwise leaves the target unchanged. This circuit can be represented as

$$C_{cnot} = \begin{pmatrix} 1 & 0 & 0 & 0 \\ 0 & 1 & 0 & 0 \\ 0 & 0 & 0 & 1 \\ 0 & 0 & 1 & 0 \end{pmatrix} \tag{2.12}$$

It is straightforward to verify, for example, that $C_{cnot} |10\rangle \rightarrow |11\rangle$.

The gate model.   The quantum gate model of computation is developed around the concept of a register Q of n qubits that is operated on by quantum circuits $C_1, C_2 \ldots$ working in parallel and in series. A given computation involves initializing Q to some known state, applying the circuits to induce state transformations, and then measuring Q at the end. Measurement causes a probabilistic "collapse" to a classical state, which is taken as the answer returned by the computation. Analysis of this computation requires finding a bound on both the time required and the probability that the answer is correct.

Because of superposition, Q can hold all 2^n states simultaneously; because of entanglement, a quantum circuit can act on all n bits (and therefore on all $N = 2^n$ states) in constant time. In contrast, a classical computing model such as a Turing machine would have to update 2^n registers to achieve the same manipulation of the state of classical R. This so-called *quantum parallelism* is the main source of optimism about the possibility of enormous speedups in computation time over classical computers.

Quantum dynamical systems.   Now consider a scenario in which the n qubits in register Q are particles in a quantum dynamical system that evolves over time according to certain forces acting on it. The (superposition) state of Q at time t is $|\Phi_t\rangle$ defined as in (2.10).

Some forces are from external sources and some arise from interactions among qubits (including entanglement). At any time t, these forces are completely characterized by a time-varying *Hamiltonian* $\mathcal{H}_t$. Mathematically $\mathcal{H}_t$ is an $N \times N$ *Hermitian matrix*, which means it has certain properties relating to symmetry and decomposability; one property is that all its eigenvalues are real. See Nielsen and Chuang [75] or Rieffel and Polak [88] for more technical development.

Every observable state $s = (0|1)^n$ has an associated *energy*, which is a real scalar that depends on $\mathcal{H}_t$. The *energy spectrum* is the set of all possible energies in the system, of size at most N. Two or more states with the same energy are said to be *degenerate*. The *ground state* s_g is an observable state having minimum energy over all basis states. A state that is not the ground state is said to be *excited*. The *first excited state* has minimum energy among all excited states—that is, the second-lowest energy in the system. Notationally the Hamiltonian serves a dual role in this framework:

- It can be considered an *operator* that changes the state of the system: $\mathcal{H}_t |\Phi_t\rangle \rightarrow |\Phi_{t'}\rangle$. The uncanny resemblance between Hamiltonians and the quantum circuits described in the previous section is not accidental. However note that Hamiltonians work in continuous rather than discrete time.

- The eigenstates of $\mathcal{H}_t$ correspond to the observable states of the system, and their eigenvalues to the energies of those eigenstates. Thus, if $|\Phi_t\rangle$ is an eigenstate we can measure its energy λ since $\mathcal{H}_t |\Phi_t\rangle \rightarrow \lambda |\Phi_t\rangle$.

Despite the notational similarities, the *acts* of applying an operator and of measuring qubit energies are not exactly the same in a real quantum computation. The implications of using Hamiltonians in these two different ways are not fully captured in the simplified notation developed here.

Thus ends our quick survey of the main notational conventions of quantum computation and quantum mechanics. The next section shows how to combine these elements to develop algorithms in the AQC model.

2.2   COMPONENTS OF AQC ALGORITHMS

AQC algorithms are designed to solve optimization problems formulated as follows: Given an *objective function* $f : \mathbb{D}^n \to \mathbb{R}$ defined on n variables $x = x_1 \ldots x_n$ from some discrete domain $\mathbb{D}$, find an assignment of values to x that minimizes $f(x)$. Sometimes the problem definition includes constraints that identify certain combinations of assignments as *infeasible* (invalid); for now we consider only unconstrained problems.

It is usually not difficult to cast a given decision problem into an optimization framework. Here are some examples.

- CNF-SATISFIABILITY: You are given a boolean expression written in conjunctive normal form, containing n variables and m clauses: $B(x) = \bigwedge_{c=1\ldots m} C_c$. For a given assignment x, let the clause function $f_c(x) = 0$ if clause c is satisfied by assignment x, and otherwise $f_c(x) \geq 1$. The problem is to find an assignment to x that minimizes $f_{cnf}(x) = \sum_c f_c(x)$.

- CIRCUIT f: You are given a logical function $\ell(b)$ defined on n bits $b = b_1 \ldots b_n$. Let the function $f_\ell(b) = 0$ if $\ell(b) = true$ and otherwise let $f_\ell(b) = 1$. Find an assignment to b that minimizes $f_\ell(b)$.

- FACTORING: Given an integer N, find a pair of integers $x, y > 1$ to minimize $f_N(x, y) = (N - xy)^2$.

An AQC algorithm to solve optimization problem P with objective function $f(x)$ works on a register Q of n qubits having superposition state $|\Phi_t\rangle$ at time t. The algorithm is described by a time-varying Hamiltonian $\mathcal{H}(t)$ specified by three components:

1. An *initial Hamiltonian* $\mathcal{H}_I$ chosen so that the ground state of the system is easy to find.

2. A *final (also called problem) Hamiltonian* $\mathcal{H}_F$ that encodes the objective function so that the ground state of Q is an eigenstate of $\mathcal{H}_F$ having minimum eigenvalue. That is, a ground state s_g of Q corresponds to an optimal solution to P.

3. An *adiabatic evolution path*, a function $s(t)$ that decreases from 1 to 0 as $t : 0 \to t_f$, for some elapsed time t_f. For now we use a simple linear path $s(t) = 1 - t/t_f$.

The Hamiltonian $\mathcal{H}(t)$ creates a gradual transition from $\mathcal{H}_I$ to $\mathcal{H}_F$ according to

$$\mathcal{H}(t) \;=\; s(t)\mathcal{H}_I + (1 - s(t))\mathcal{H}_F. \tag{2.13}$$

This Hamiltonian is an AQC algorithm for solving P.

2.2.1   A SIMPLE EXAMPLE

We follow Farhi et al. [35] to describe an AQC algorithm for the *2-bit Disagree* problem, which is to output $|01\rangle$ or $|10\rangle$ but not $|00\rangle$ or $|11\rangle$. The objective function on bits $x = x_1 x_2$ is

$$f_{dis}(x) \;=\; \begin{cases} 0 & \text{if } x_1 \neq x_2, \\ 1 & \text{otherwise.} \end{cases} \tag{2.14}$$

The final Hamiltonian.   The goal is to build a Hamiltonian such that each eigenstate matching an assignment to x has an eigenvalue that matches $f_{dis}(x)$. Recalling the Pauli-z operator from (2.4), we set the final Hamiltonian to be

$$\mathcal{H}_{dis} \;=\; \frac{1 + \sigma_1^z \sigma_2^z}{2} \tag{2.15}$$

where the subscript i on σ_i^z indicates that the operator is applied only to the individual qubit ϕ_i in $|\Phi\rangle = |\phi_1 \phi_2\rangle$. In general, to build a full $N \times N$ Hamiltonian from smaller 2×2 Pauli operators that are applied to individual qubits, we use the tensor product (2.8) to combine a sequence of Identity and Pauli operators:

$$\sigma_i^z \sigma_j^z \;=\; I \otimes \cdots \otimes \sigma_i^z \otimes \cdots \otimes \sigma_j^z \otimes \cdots \otimes I, \tag{2.16}$$

where the σ^z operators are in positions i and j. Application of the Hamiltonian (2.15) to Q yields the following:

$$\mathcal{H}_{dis}\,|00\rangle \;\rightarrow\; 1\,|00\rangle \quad \text{via} \quad \frac{1 + (+1)(+1)}{2}$$

$$\mathcal{H}_{dis}\,|01\rangle \;\rightarrow\; 0\,|01\rangle \quad \text{via} \quad \frac{1 + (+1)(-1)}{2}$$

$$\mathcal{H}_{dis}\,|10\rangle \;\rightarrow\; 0\,|10\rangle \quad \text{via} \quad \frac{1 + (-1)(+1)}{2}$$

$$\mathcal{H}_{dis}\,|11\rangle \;\rightarrow\; 1\,|11\rangle \quad \text{via} \quad \frac{1 + (-1)(-1)}{2}. \tag{2.17}$$

That is, the eigenstates and eigenvalues of $\mathcal{H}_{dis}$ correspond exactly to $f_{dis}(x)$. Eigenvalues are energies, so the ground states of this system are $|01\rangle$ and $|10\rangle$.

The Initial Hamiltonian.   To form the initial Hamiltonian we use the Pauli-x operator σ^x from (2.3). The eigenvectors of σ^x are $|x_+\rangle$ and $|x_-\rangle$ as in (2.6). We construct $\mathcal{H}_I$ according to

$$\mathcal{H}_I = \sum_{i=1}^{2} \frac{1}{2}(1 - \sigma_i^x). \tag{2.18}$$

As before the notation σ_i^x means that the operator is applied only to qubit i. Recalling (2.7) and applying this Hamiltonian to $|\Phi\rangle$ we obtain

$$\mathcal{H}_I |x_+x_+\rangle \quad \rightarrow \quad 0 |x_+x_+\rangle$$

$$\mathcal{H}_I |x_+x_-\rangle \quad \rightarrow \quad 1 |x_+x_-\rangle$$

$$\mathcal{H}_I |x_-x_+\rangle \quad \rightarrow \quad 1 |x_-x_+\rangle$$

$$\mathcal{H}_I |x_-x_-\rangle \quad \rightarrow \quad 2 |x_-x_-\rangle, \tag{2.19}$$

so the ground state of $\mathcal{H}_I$ is $|x_+x_+\rangle$. Re-writing $|x_+x_+\rangle$ in terms of the standard basis we have

$$|x_+x_-\rangle \quad = \quad \frac{1}{4} |00\rangle + \frac{1}{4} |01\rangle + \frac{1}{4} |01\rangle + \frac{1}{4} |11\rangle . \tag{2.20}$$

This means that $\mathcal{H}_I$ creates a ground state for which each eigenstate in the standard basis is equally likely to be observed. As a general rule in AQC, the initial Hamiltonian is specified in a basis that is orthogonal to that of the final Hamiltonian, so as to obtain an equiprobable superposition like this one. Using an orthogonal basis also ensures that a type of bad outcome called a "level crossing" (described below) is unlikely to occur.

The adiabatic evolution path.   We use the linear transition $s(t) = 1 - t/t_f$.

The algorithm.   Our AQC algorithm to solve the Two-Bit Disagree problem is now completely specified by

$$\mathcal{H}(t) = s(t)\, \mathcal{H}_I + (1 - s(t))\, \mathcal{H}_{dis}. \tag{2.21}$$

To run this algorithm, acquire two qubits $Q = (q_1, q_2)$ having states described by $|\Phi_t\rangle$ at time t. Apply energies to the qubits according to $\mathcal{H}(t)$ for $t : 0 \rightarrow t_f$. At the end of the computation, register Q is in a low-energy superposition state $|\Phi_T\rangle$ according to $\mathcal{H}_{dis}$. Reading Q in the standard basis causes a probabilistic "collapse" to a binary string that is taken as the solution x. According to the adiabatic theorem, if certain conditions hold, the algorithm will almost surely return an optimal solution, either 01 or 10.

2.2.2   THE ADIABATIC THEOREM

Formal development of the adiabatic theorem is outside the scope of this discussion. Here we simply state the theorem in a way that can be used to bound the computation time of an AQC algorithm. Several formulations have appeared: see Aharonov et al. [1], Childs [22], or Reichardt [89] for discussions and proofs. The development below follows Van Dam et al. [105].

The adiabatic theorem was formulated by Born and Fock [16] to describe certain properties of quantum particle processes, which evolve according to the *Schrödinger equation*,

$$i\hbar \frac{d}{dt}|\Phi_t\rangle \;=\; \mathcal{H}(t)|\Phi_t\rangle. \tag{2.22}$$

Here i is the imaginary unit and $\hbar = h/2\pi$ where h is Planck's constant ($\approx 6.63 \times 10^{-34}$ Joule-second). This equation says that the instantaneous rate of change of the process (the partial derivative of the state transition vector) is proportional to the energy in the process (on the right hand side of the equation).

Suppose $\mathcal{H}(t)$ describes an AQC algorithm acting on Q, which evolves over time $t : 0 \to t_f$. Let $s(t)$ be strictly decreasing from 1 to 0 as $t : 0 \to t_f$. Let $\widetilde{\mathcal{H}}(s)$ be the equivalent Hamiltonian in the time scale $s : 0 \to 1$, which evolves at a *rate* determined by t_f. Let $\tau(s)$ determine the rate at which the Hamiltonian changes as a function of s. Now we have

$$\frac{d}{ds}|\Phi_s\rangle \;=\; -i\tau(s)\widetilde{\mathcal{H}}(s)|\Phi_s\rangle. \tag{2.23}$$

Assume that $\widetilde{\mathcal{H}}_s$ has a nondegenerate ground state throughout time s. Let δ_s denote the *spectral gap*, the (positive) difference between the eigenvalues of the ground state and of the first excited state, at any time s. Let δ_m be the *minimum spectral gap* $\delta_m = \min_s(\delta_s)$. Now suppose $\tau(s)$ is bounded by

$$\tau(s) \;\gg\; \frac{\left\| \frac{d}{ds}\widetilde{\mathcal{H}}(s) \right\|}{(\delta_m)^2}, \tag{2.24}$$

where $\|\cdot\|$ is the matrix norm induced by the L_2 metric.

Let $|\Phi_g\rangle$ denote the ground state of the final Hamiltonian. The adiabatic theorem states that:

1. If Q is in ground state at $s = 0$, and

2. if δ_m is strictly greater than 0 throughout time s, and

3. if the process evolves slowly enough to obey (2.24),

then the process will finish in ground state ($|\Phi_{t_f} = \Phi_g\rangle$) with sufficiently high probability. As a general rule it is easy to show that the numerator in (2.24) is bounded above by a low-degree polynomial in n. Therefore, computation time t_f is polynomial in n exactly when the minimum spectral gap δ_m is inverse polynomial in n.

The spectral gap.   Panel (a) of Figure 2.2 shows four eigenvalues for a hypothetical algorithm, moving in time s. The spectral gap δ_m is small at mid-transition. In this case, a too-quick transition time t_f—violating (2.24)—yields an increased probability that the system will "jump" from ground state to the first excited state and be carried away to a non-optimal solution at the end of the transition. Panel (b), from Farhi et al. [36], shows how the four eigenvalues of our Two-Bit Disagree algorithm move over time s. The spectral gap δ_s is large throughout, except at the very end when the two ground states come together. This is not a problem since we do not care which of the two states is returned.

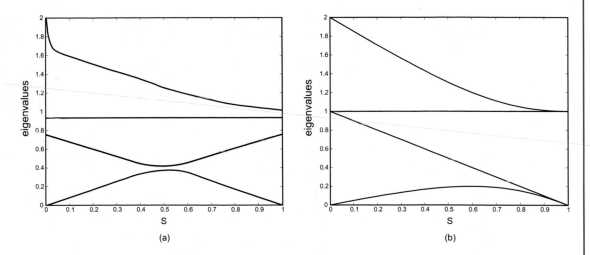

Figure 2.2: Panel (a) shows the transition over time of four eigenvalues for a hypothetical algorithm with a small gap δ_m at mid-transition. Panel (b) shows the four eigenvalues for the 2-Bit Disagree algorithm.

A *level crossing* corresponds to a situation where $\delta_m = 0$ and therefore the adiabatic theorem does not apply. One reason for choosing $\mathcal{H}_I$ to operate in a basis that is orthogonal to that of $\mathcal{H}_F$ is that level crossings are highly unlikely to occur.

The next section presents a more interesting algorithm and surveys some approaches to analyzing computation time.

2.3   AN ALGORITHM FOR EXACT COVER

In this section we present the AQC algorithm described by Farhi et al. [36] for Exact Cover (EC), cast as a variation on 3-Satisfiability: You are given a boolean expression $B(x) = \wedge_{c=1}^m C_c$ defined on n binary variables $x = x_1 \ldots x_n$ where $x_i \in \{0, 1\}$. The expression contains m clauses

C_c, each containing three variables labeled (x_{c1}, x_{c2}, x_{c3}), where each variable can equal x_i or $\neg x_i = (1 - x_i)$.[1]

A clause is satisfied if exactly one of its variables has the value 1. We define the objective function $f_{ec}(x)$ as a sum of individual clause functions $f_c(x)$, as follows.

$$f_c(x) = (1 - x_{c1} - x_{c2} - x_{c3})^2$$

$$f_{ec}(x) = \sum_c f_c(x). \tag{2.25}$$

Note this objective function is always nonnegative, and is equal to 0 exactly when all clauses are satisfied. Multiplying out and collecting terms, we obtain

$$f_{ec}(x) = -2m - \sum_i B_i x_i + \sum_i B_i x_i x_i + \sum_{i<j} C_{ij} x_i x_j$$

$$= -2m + \sum_{i<j} C_{ij} x_i x_j, \tag{2.26}$$

where B_i counts the total number of times x_i appears in any clause, and C_{ij} counts the number of times the pair x_i, x_j appears together in a clause.

Final Hamiltonian.   To define the final Hamiltonian we apply $x_i = (1 - s_i)/2$ to replace binary variables $x_i \in \{0, 1\}$ with spin values $s_i \in \{-1, +1\}$, to obtain an energy function:

$$\hat{f}_{ec}(s) = -2m + \sum_{i<j} C_{ij} \frac{(1 - s_i)(1 - s_j)}{4}. \tag{2.27}$$

This energy function corresponds to the eigenvalues of a final Hamiltonian and is minimized by:

$$\mathcal{H}_{ec} = \sum_{i<j} C_{ij}(1 - \sigma_i^z)(1 - \sigma_j^z). \tag{2.28}$$

Initial Hamiltonian.   For the initial Hamiltonian we use

$$\mathcal{H}_I = \sum_{i=1}^m \frac{1 - \sigma_i^x}{2}. \tag{2.29}$$

As before this Hamiltonian using the Pauli-x operator places the qubits in equal superposition in the z-basis so that

$$|\Phi_B\rangle = \frac{1}{2^{n/2}} \sum_1 \sum_1 \cdots \sum_n |\phi_1\rangle |\phi_2\rangle \cdots |\phi_n\rangle. \tag{2.30}$$

[1]Notice that binary variables $\{0, 1\}$ are used in expressions that mix boolean and arithmetical operators, with the standard interpretation $0 = False$ and $1 = True$.

Adiabatic path.   We again use the simple linear path

$$s(t) = 1 - t/t_f.$$

Our AQC algorithm for EC is

$$\mathcal{H}(t) \quad = \quad s(t)\mathcal{H}_I + (1 - s(t))\mathcal{H}_{ec}. \tag{2.31}$$

2.3.1   RUNTIME ANALYSIS

In order to obtain an upper bound on computation time t_f as a function of problem size n, we need an upper bound on the numerator $\mathbb{H} = \left\| \frac{d}{ds} \widetilde{\mathcal{H}}(s) \right\|$ and a lower bound on the denominator $(\delta_m)^2$ from (2.24).

$\mathbb{H}$ is bounded above by the largest difference in eigenvalues obtainable from the Hamiltonians $\mathcal{H}_F$ and $\mathcal{H}_I$. The *spectrum* of a Hamiltonian is the set of its possible eigenvalues. In this case the spectrum of $\mathcal{H}_F$ is $[0 \dots 6m]$ (based on the maximum possible value of C_{ij}), and the spectrum of $\mathcal{H}_I$ is $[1 \dots m]$. So we have $\mathbb{H} \in O(m) = O(n^3)$ assuming no redundant clauses.

As a general rule, analysis of the minimum spectral gap δ_m for a given algorithm $\mathcal{H}(s)$ and input instance (or input class) I is quite difficult. In particular it is not known whether δ_m is polynomial or exponential for the algorithm of (2.31). Indeed, since EC is NP-Complete, finding an exponential lower bound would be tantamount to separating P and NP in this computational model.

The original AQC algorithm described by Farhi et al. [35] is a generalization of (2.31) to generic 3-Satisfiability problems (G3-SAT) where each clause is an *arbitrary* boolean expression on three variables. Their development in fact creates a family of algorithms, each described by a tuple $(\mathcal{H}_I, \mathcal{H}_F, s(t), t_f)$. We finish this section with a quick survey of what is known about some members of this family.

Note first that empirical evaluation of satisfiability algorithms often focuses on a problem category consisting of random instances generated at the so-called *critical point*. These instances are created by random assignment of n variables (or their negations) to m clauses, with $r = m/n$; in k-SAT formulations each clause contains k variables. It has been widely observed that random instances in this model undergo a phase transition as r moves over a critical point r_0, changing from almost surely satisfiable when $r < r_0$ to almost surely unsatisfiable when $r > r_0$. Furthermore (see for example Cheeseman et al. [20] or Mitchell et al. [72]), many satisfiability algorithms run into serious difficulties (long computation times) when presented with random instances generated at the critical point r_0. For 3CNF-SAT, it has been empirically observed that r_0 is near 4.26. The phase transition appears in other k-SAT formulations, although the location of r_0 varies.

- Farhi et al. [35] have obtained polynomial upper bounds on their G3-SAT algorithm when restricted to simple instances that can also be solved classically in polynomial time.

- Farhi et al. [36] describe numerical calculations suggesting polynomial solution times for EC problems on random instances of size up to $n = 20$, generated near the critical point.

They observe that median time to achieve a fixed success probability closely fits a quadratic curve; however these problem sizes are too small to support reliable extrapolation.

- Van Dam et al. [2001] describe example instances for which the algorithm of Farhi et al. [35] has δ_m shrinking exponentially in n, implying an exponential lower bound on computation time. However Farhi et al. [34] show that the small gap in that construction can be avoided by replacing the simple linear adiabatic path $s(t) = 1 - t/t_f$ by an alternative path. With this modification the exponential lower bound no longer applies.

- Hogg [51] describes numerical simulations of an AQC algorithm to solve 3CNF-SAT instances generated at the critical point and screened to be satisfiable. He observes that solution times scale better than those for *GSAT*, a well-known classical heuristic solver, on problem sizes up to $n = 30$.

- Schaller and Schützhold [96] observe that the occurrence of small gaps during the anneal process appear to be associated with a type of phase transition marked by rapid changes in the ground state. They adopt different $\mathcal{H}_F$ and $\mathcal{H}_I$ from those in Farhi et al. [34], that are intended to mitigate that situation. Numerical simulations on random EC instances up to size $n = 24$ suggest that their alternative algorithm converges more quickly than the original.

- Young et al. [110] present results of Monte Carlo simulations using random satisfiability instances of size $n \leq 128$. They observe that in the median case the minimum spectral gap δ_m appears to grow inversely as $O(n^2)$, which would suggest an $O(n^4)$ median computation time for this instance class.

- Choi [26] describes two final Hamiltonians for the EC problem: one has an exponentially small gap for a given class of problem instances (the proof is by Altshuler et al. [2]); the second incorporates an extra term for which the lower bound proof does not apply. She also describes a simple modification to the Hamiltonian that appears to be robust against this lower bounding argument.

- Choi [26] also points out that solving a decision problem requires only that the final Hamiltonian and the objective function share the same *minimum values*, not that they match at all points in the solution space. This opens up considerable flexibility in algorithm design.

- Farhi et al. [34] describe a class of final Hamiltonians that induce small gaps that shrink exponentially in n with respect to their AQC algorithm. Then they show how to remove those small gaps by introducing a third Hamiltonian to the algorithm.

- Farhi et al. [34] also describe simulation experiments using specially contrived G3-SAT instances, for which their algorithm does not appear to be efficient. However running the algorithm repeatedly with randomly selected adiabatic paths appears to yield polynomial solution times on average.

- Hen and Young [50] use a simulation of an AQC algorithm to study performance on three types of SAT problems—locked 1-in-3 SAT, locked 2-in-4 SAT, and 3-regular 3-XORSAT (the last one is in P). They observe exponential solution costs for this algorithm.

And so it goes. Mirroring the progress of algorithm development in classical computation, we see new algorithms proposed, leading to new analyses, causing some algorithms to be discarded and others—faster and more robust—to be proposed. In the AQC paradigm, the runtime analysis of a given algorithm can change if any one of its four components—the final Hamiltonian, the initial Hamiltonian, the shape of the adiabatic path, or the transition time—is modified.

So far these analyses have not settled the larger questions about P vs. NP, nor whether quantum computation yields a quantum speedup over classical computation for the satisfiability problem. However they do provide valuable intuitions about what makes AQC algorithms efficient. Furthermore, some general strategies for algorithm design have emerged:

- Somma and Boixo [101] describe a technique for *spectral gap amplification* by which the final Hamiltonian can be modified to achieve speedups in some cases. They also describe a family of Hamiltonians for which spectral gap amplification is not possible.

- Dickson and Amin [30] present a simple procedure that can be used to avoid small gaps by iterating a given AQC algorithm: after each iteration, information about the solution sample is used to adjust the final Hamiltonian away from those dangerous areas.

2.4   COMPLEXITY CLASSES

Figure 2.3 shows the major complexity classes relating to this area. The classes outlined in black and red refer to decision problems, where the outcome of the computation is a binary value, yes or no. The classes in blue on the right side refer to optimization problems, where the outcome is a minimum-cost solution, and the class on the left denotes *counting problems* where the output is a count (or a list) of minimum-cost solutions. The classes in red correspond to probabilistic and quantum models of computation. As is famously the case with P vs. NP, it is not known which of these inclusions are proper.

Here are brief descriptions of the decision problem classes.

- **P, NP, PSPACE**, and **EXPTIME** are the familiar classes for deterministic and nondeterministic computation. **NPC** is the set of NP-Complete problems.

- **BPP** is bounded-error probabilistic polynomial time. A probabilistic computer is allowed to toss random coins and to return an answer that is correct with at least some fixed probability p, often taken to be 2/3. This class of problems can be solved by a probabilistic algorithm in polynomial time, with an error probability of at most 1/3.

- **MA**, which stands for Merlin-Arthur computation, is the probabilistic analog of NP. Merlin is computationally unbounded and provides a certificate that Arthur can verify using a BPP machine.

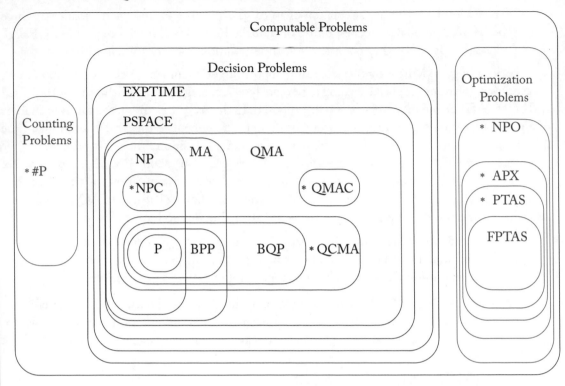

Figure 2.3: Classical and quantum complexity classes. The classes in black and blue are defined on classical models of computation. The classes in red are defined on probabilistic and quantum models of computation. The stars mark classes that are also NP-hard.

- **BQP** is bounded-error quantum polynomial time. A quantum model of computation uses an arrangement of quantum circuits (in the gate model) operating on qubits. These problems can be solved by a quantum computer in polynomial time with an error probability of at most 1/3.

- **QMA** stands for Quantum Merlin-Arthur. This is the set of decision problems for which yes-instances can be verified by a quantum computer in polynomial time, and no-instances rejected in polynomial time, with error probability at most 1/3. This is the quantum analog of NP, related to BQP as NP is related to P. **QMAC** is the set of complete problems for this class.

- **QCMA** stands for Quantum-classical Merlin-Arthur. This class is similar to QMA but the witness for a yes-instance must be a classical string.

For the problem classes shown in blue, a problem P is defined on a set of instances $\mathcal{I}$ from a discrete domain, by specifying: the set $F \subset \mathcal{I}$ of feasible solutions; an objective function $f(x)$: $\mathcal{I} \to \mathbb{R}$; and whether the objective function is to be minimized or maximized. The optimization problem is to find a feasible solution that minimizes (resp. maximizes) the objective function. The counting problem is to return a count of all optimal solutions. (We assume a minimization problem henceforth.)

Optimization problems can be considered in terms of their *approximation complexity*: that is, given an instance $I \in P$ for a minimization problem with optimal solution s^*, how difficult is it to return a feasible solution s within some difference $\Delta = |f(s) - f(s^*)|$ or ratio bound $r \leq f(s)/f(s^*)$? If an algorithm returns feasible solutions such that $\Delta = 0$ we say it is an *exact algorithm*. If it returns solutions for which a non-trivial (usually constant or low-polynomial) bound on Δ or r is known analytically, we call it an *approximation algorithm*. Some complexity classes for optimization problems are shown in blue in Figure 2.3.

- **NPO** is the class of optimization problems that have a decision version in NP.

- **APX** is the set of optimization problems that can be approximated within a constant factor r in polynomial time. Examples include Minimum Vertex Cover, Maximum 3-Satisfiability, and Bin Packing. It is also known that the Traveling Salesman Problem is not in APX.

- **PTAS** is the set of problems for which there exists a polynomial-time approximation scheme. This is a parameterized algorithm A_r that for any rational $r > 1$, returns a solution cost within r of optimal, in time polynomial in problem size n, but inversely exponential in r. An example in this class is Maximum Independent Set for planar graphs.

- **FPTAS**. A fully polynomial time approximation scheme is a parameterized approximation algorithm that can return a solution bounded by r in time that is polynomial in both n and $1/(r - 1)$.

- **#P** (pronounced sharp-P) is the set of counting problems that have decision analogs in NP and optimization analogs in NPO. Given an optimization problem instance I the algorithm must return a count of the number of optimal solutions to I.

NP-hard problems.   The stars (*) in Fig 2.3 mark *NP-hard* problem classes. A problem A is NP-hard if there exists a polynomial-time reduction from an NP-complete problem B, to A. Such a reduction implies that an algorithm to solve A in polynomial time can be used as a subroutine to solve B—and therefore all NP-complete problems—in polynomial time. (Note a polynomial time algorithm for B does not necessarily imply the existence of a polynomial time algorithm for A.) We have, for example, $NPC = NPH \cap NP$ and $QMAC = NPH \cap QMA$.

Any decision problem in NP can be solved by an *exact* algorithm for the analogous problem in NPO, since an exact algorithm returns a solution that provides the witness for the verification step. It is known that no NP-hard optimization problem can belong to FPTAS unless $P = NP$ (see Jansen [54]).

2.4.1   AQC AND RELATED MODELS

Where does AQC fit into the complexity classes of Figure 2.3? The main result in this area is a proof by Aharonov et al. [1] that a universal version of AQC can simulate any computation in the quantum gate model (QGC) with at most polynomial slowdown. Combined with the result by Farhi et al. [36] that the quantum gate model can simulate any AQC computation, this implies that AQC is polynomially equivalent to QGC. Therefore AQC—which contains the set of computable optimization problems—must also contain complexity classes analogous to QMA and BQP, that contain NPO and FPTAS respectively.

Van Dam et al. [105] have shown that an AQC algorithm for Grover's problem—searching in an unordered set—exhibits quadratic speedup over classical algorithms, as is also the case with the gate model.

Locality.   The Aharonov et al. [1] proof assumes a computational model in which the initial and final Hamiltonians $\mathcal{H}_I$ and $\mathcal{H}_F$ are *k-local*. A k-local Hamiltonian can be written as the sum of independent Hamiltonians that each work on at most k qubits. For example, the EC algorithm (2.31) is 2-local because $\mathcal{H}_F$ is the sum of operators involving at most two qubits.

Local Hamiltonians are considered better candidates for realization on a physical quantum computer which may have limited connectivity. Sometimes locality is defined according to topology: for example, the final Hamiltonian may be restricted to involve only interactions among neighboring qubits in a grid.

Aharonov et al. [1] show that any quantum circuit C_f using at most L gates can be simulated by an adabatic quantum computer using one of these models:

- Five-local Hamiltonians, with at most $O(L^5)$ expansion in problem size.

- Three-local Hamiltonians, with at most $O(L^{14})$ problem overhead.

- Two-local Hamiltonians interacting on a 2D grid of qubit cells, assuming each cell can represent six different states.

Restrictions on the universal model.   Published AQC algorithms use final Hamiltonians with real-valued diagonal elements that exactly match the objective functions $f(x)$ and with off-diagonals equal to zero. This category of Hamiltonian $\mathcal{H}_F$ creates a classical ground state (with no superposition) at the time the solution is read. A much wider choice of final Hamiltonians is available in the general computational model; indeed Aharonov et al. [1] remark that their proof of equivalence with the gate model assumes that the final Hamiltonian is not restricted to be classical in this way.

Bravyi et al. [18] define a *stoquastic* Hamiltonian as one for which off-diagonal elements are real and non-positive. They show that the problem of minimizing local stoquastic Hamiltonians is hard for the class AM (similar to MA, a probabilistic version of NP with two rounds of communication between the prover and the verifier).

Biamonte and Love [13] consider restrictions to the universal AQC model with an eye toward realizability in physical machines. They describe two categories of Hamiltonians that are sufficient to achieve QMA-complete computation: this means the models can solve NP-hard problems. They also describe how to modify these models to achieve universal AQC computation.

The quantum annealing chips built by D-Wave require problem Hamiltonians specified with reals on the diagonal and zeros on the off-diagonal, and have a connectivity structure that is 2-local. Therefore the proof of Aharonov et al. [1] of AQC equivalence with the quantum gate model does not apply to these chips in their current design. The approximation problem implemented in the D-Wave hardware topology has a classical PTAS; see Section 4.1 for more.

CHAPTER 3

Quantum Annealing

Quantum annealing (QA) is a heuristic approach to solving problems in combinatorial optimization, originally developed for implementation on classical machines. It may be seen as a variation on the more familiar simulated annealing (SA) metaheuristic. (See Figure 3.1 for a definition of this term.) The idea of incorporating a model of quantum rather than thermal annealing into a heuristic optimization framework appears to have been independently suggested by several researchers. Early work may be found, for example, in Apolloni et al. [3], Ray et al. [87], Tirumalai et al. [104], Finnila et al. [38], and Kadowaki and Nishimori [56].

It is now understood that quantum annealing algorithms can be developed using the "basic instruction set" of AQC as well as that of classical computers. Thus QA provides a conceptual bridge between adiabatic quantum computation and classical optimization, which (judging from the number of algorithms proposed in the last few years) has served to catalyze the algorithm design process.

Many authors now use the terms AQC and QA interchangeably. Here, however, we observe a distinction suggested by Rose and Macready [92] that applies not so much to the algorithm as to how it is analyzed:

- The term *AQC algorithm* refers to an algorithm in the universal AQC model. AQC algorithms can be designed to solve any Turing-computable problem; they can simulate quantum algorithms in the gate model with at most polynomial increase in computation time.

- A *QA algorithm* is designed to solve a (typically NP-hard) combinatorial optimization problem. This implies a restriction on the final Hamiltonian $\mathcal{H}_F$ so that it represents a classical objective function, as described in Section 2.4. On the other hand certain properties of the AQC model are relaxed: for example, we do not assume that the entire computation take place in ground state.

This chapter considers algorithms designed in the QA framework, focusing on applications in combinatorial optimization. The analysis of any algorithm depends on the platform on which it runs; in the case of QA, we have three options:

- The algorithm may run on a theoretically ideal AQC platform (analogous to a Turing machine) in an *adiabatically closed system*, which means that it experiences no interference from the environment. With this assumption, the algorithm is probabilistic and the Adiabatic Theorem may be used to bound the tradeoff between computation time and the probability that the system finishes in ground state.

- A **complete** or **exact** algorithm guarantees to return a minimum-cost solution to a given minimization problem. (Respectively, a maximum-cost solution to a given maximization problem.) Assuming $P \neq NP$, such an algorithm runs in exponential worst-case time.

- An **approximation algorithm** runs in polynomial time and has a guaranteed bound on solution quality, typically expressed as a bound on the ratio ApproxSolution/OptimalSolution.

- A **probabilistic algorithm** uses an internal random source to perform its computations. There may be a known probabilistic bound on the tradeoff between computation time and solution quality.

- A **heuristic** is an algorithm that has weak or nonexistent theoretical guarantees on runtime and/or solution quality.

- A **metaheuristic** is a problem-generic solution strategy for solving NP-hard problems, on par with an algorithm paradigm like dynamic programming.

- **Heuristic search** is a type of metaheuristic. **Quantum annealing** and **simulated annealing** (also metaheuristics) are variations on heuristic search.

- In optimization, **Monte Carlo** refers to a category of metaheuristic. This use of the term is distinct from its meaning in complexity theory.

- A **hybrid metaheuristic** combines distinct metaheuristics, for example using one to control the "long range" and another for the "local" search strategies.

Figure 3.1: Categories of classical optimization algorithms.

- The algorithm may be implemented to run on a physically realized quantum platform. It is a matter of natural law that a real quantum computer must work in an *open system*: that is, it cannot be perfectly isolated from its environment. Energy from the environment—of all types, including thermal, magnetic, and radiation—creates noise that interferes with the computation and reduces the probability of finishing in ground state. Thus the theoretical guarantees of AQC do not necessarily apply. When no suitable performance bounds are known, we consider the QA algorithm to be a heuristic.

- The algorithm can be implemented to run on an everyday classical platform. A classical computer is not vulnerable to the so-called energy bath of an open system; on the other hand it does not have support for quantum features like superposition and entanglement.

Therefore a classical implementation of QA must incorporate some type of strategy to simulate or approximate a truly quantum computation. The category of algorithm (exact or heuristic) depends on the strategy.

Section 3.1 gives an overview of classical implementations of QA algorithms, which are frequently compared to their simulated annealing (SA) counterparts. (Implementations on quantum platforms are discussed in Chapter 5.) Section 3.2 surveys QA algorithms for several problem domains, including *Ising Spin Model* and *Boolean Satisfiability*, and others. The primary focus is on showing how to formulate a QA algorithm for a given problem.

This chapter covers only a few highlights of a large body of research on this algorithmic approach. For more thorough discussions of QA and its properties, see Das and Chakrabarti [27], Kadowaki and Nishimori [56], or Morita and Nishimori [74].

3.1   OPTIMIZATION AND HEURISTIC SEARCH

We start with an overview of classical implementations of simulated annealing (SA) and quantum annealing (QA). These two variations on heuristic search are similar in that both draw on metaphors from natural physical processes to solve combinatorial optimization problems. The distinction is that simulated annealing models a thermal annealing process in metallurgy, whereas quantum annealing models particle fluctuations in quantum dynamics.

Heuristic Search   Suppose we have a minimization problem P defined on n binary variables $x = x_1 \ldots x_n$. The problem statement includes an *objective function* $f(x)$ that assigns a cost to each solution $x \in \mathbb{B}^n = \{0, 1\}^n$. For now we assume that every $x \in \mathbb{B}^n$ is a feasible solution.

The *solution space* $\mathbb{D}^n$ is represented by a (possibly infinite) graph with nodes as solutions and edges defined by a *neighborhood rule*. Together, the graph and the objective function create a *landscape*, where hilltops correspond to solutions with high cost and valleys to solutions with low cost. Figure 3.2 shows how a problem landscape is typically represented.

A *local search* algorithm starts at some initial node in the space and then iterates, stepping from node to neighbor node and generally heading toward low-cost regions of the space, until terminated by some stopping rule.

There is an art to choosing a good neighborhood rule to help local search succeed. The solution space must be connected, and ideally it should have small diameter, but too much connectivity may cause the algorithm to never venture far from its starting node. We prefer landscapes that are "smooth" in $f(x)$ so that optimal solutions have low-cost neighbors: this creates a wide valley around the global optimum, which is easier to locate than a narrow chasm. Neighborhood rules that connect solutions with small incremental differences (such as low Hamming distance) are usually preferred. On the other hand, a landscape with large plateaus of nearly equal-valued neighbors is difficult to traverse because the algorithm can't tell if it is making progress.

If the landscape is smooth and generally sloping downward toward the optimal solution, a simple *greedy-descent* approach that always steps to the least-cost neighbor works well. But if

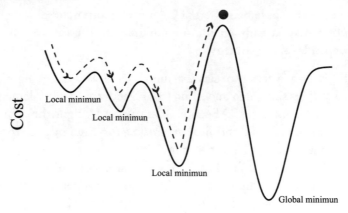

Figure 3.2: The objective function and neighborhood rule define a solution landscape. A heuristic search algorithm steps from node to neighbor node looking for an optimal solution while avoiding getting stuck in local minima.

the landscape contains many local minima, the greedy approach can get stuck, possibly move in circles, and never find the optimal solution.

A huge number of mechanisms have been proposed for moving the search out of local minima and toward low-cost regions of the solution landscape. These are generally called *heuristic search* methods; see Edelcamp and Schrödl [33] or Michalewicz and Fogel [73] for an introduction. Simulated annealing and quantum annealing are two types of heuristic search.

```
Simulated Annealing
 1  x = initial state,  c = f(x)
 2  iter = 1
 3  while !stoppingRule():
 4      T = coolingSchedule(iter++)
 5      xnew = neighborOf(x)
 6      cnew = f(xnew)
 7      if cnew < c:
 8              x=xnew,  c=cnew
 9      else if runif() > AcceptProb(T, cnew-c ):
10              x=xnew, c=cnew
11 return (x)
```

Figure 3.3: The Simulated Annealing (SA) metaheuristic.

Simulated annealing.   The simulated annealing strategy, proposed by Kirkpatrick et al. [61], is sketched in Figure 3.3. The algorithm starts at an initial state x having cost c, and then iterates to walk through the problem landscape. At each iteration, it selects a random neighbor of the current solution (line 5). If the new solution has lower cost, it becomes the current solution (lines 7, 8) and we say the downhill move is accepted. If the new solution has higher cost, the uphill move is accepted according to a probability that depends on the *temperature T* and on the difference between costs cnew and c (line 9). Letting $\Delta = cnew - c$, this probability is often defined as

$$Prob[Accept(T, \Delta)] \quad = \quad \min\left(1, e^{-\Delta/T}\right) \tag{3.1}$$

where e is the base of the natural logarithm. The temperature parameter T is initialized at some high value and then decremented at each iteration according to a *cooling schedule* (line 4). Many variations on this general strategy may be considered. A common approach is to initialize T to a value near $E[|cnew - c|]$, decrement according to the schedule $T = a/\ln(i)$ for iteration i and some constant a, and stop when T is below some convenient threshold near 0.

Simulated annealing can be viewed as taking a random walk in the problem space according to a Markov chain parameterized by T. Geman and Geman [46] show that under certain conditions, the random walk is guaranteed to reach a steady state distribution such that the probability π of observing state x at a fixed temperature T is given by

$$\pi_T(x) \quad = \quad \frac{1}{\alpha_T} e^{\left(\frac{-f(x)}{T}\right)}, \tag{3.2}$$

where α_T is a normalizing constant and $f(x)$ is our objective function. Therefore, as $T \to 0$ the probability density becomes concentrated on the global optimum, but the rate of convergence to steady state becomes extremely slow. The rationale behind simulated annealing is to accelerate convergence to a steady state concentrated at low-cost solutions by gradually decreasing T over time. If the temperature $T(i)$ is reduced gradually enough, that is, if

$$T(i) \quad \geq \quad \frac{cn}{\log i}, \tag{3.3}$$

for constant c, then SA is guaranteed to find the optimal solution with probability approaching 1 as $i \to \infty$.

Despite this pessimistic worst-case bound on convergence, SA can be quite successful in practice. Many variations on simulated annealing have been implemented and evaluated experimentally across many problem domains. Very roughly, it is considered a successful strategy when little *a priori* knowledge is available as to the structure of the solution landscape, but which can be outperformed by solvers that apply domain-specific knowledge. Much depends on matching the heuristic parameters—neighborhood rule, starting temperature, cooling schedule, total iterations, etc.—to properties of the inputs at hand. See van Laarhoven and Aarts [62], Bertsimas and Tsitsiklis [12], or Michalewicz and Fogel [73] for more about simulated annealing.

Quantum annealing.   Like simulated annealing, quantum annealing can be viewed as a type of heuristic search that moves over a solution landscape, looking for low-lying areas: that is, QA carries out a random walk described by a parameterized Markov chain. However, instead of using a temperature parameter T to control the probability of taking an uphill step at each iteration, QA uses a *transverse field coefficient* Γ (also called the tunneling coefficient) to control the traversibility of the solution landscape. Like T, Γ starts at a high value and is gradually decremented over time according to a given schedule.

Recall from Chapter 2 that our objective function $f(x)$ can be represented by a final Hamiltonian $\mathcal{H}_F$, which is an $N \times N$ Hermitian matrix with costs $f(x)$ on the diagonal and zeros elsewhere.

Quantum annealing introduces a *transverse field Hamiltonian* (also called the disordering Hamiltonian) $\mathcal{H}_D$ that does not commute with $\mathcal{H}_F$. The new Hamiltonian is scaled by $\Gamma(t)$, which is initialized to some high value and gradually reduced to 0 over time. This creates a new time-dependent Hamiltonian,

$$\mathcal{H}(t) \;\; = \;\; \mathcal{H}_F + \Gamma(t)\mathcal{H}_D. \tag{3.4}$$

The disordering Hamiltonian $\mathcal{H}_D$ introduces kinetic energy to the annealing process in the form of quantum fluctuations of the solution space. Decreasing $\Gamma(t)$ moves the system closer to $\mathcal{H}_F$ while dampening the quantum fluctuations.

One way to visualize this idea is to imagine the random walk taking place in a problem landscape that changes shape over time. When $\Gamma(t)$ is large, the second term dominates and the landscape is "disordered" relative to the objective function $f(x)$; when $\Gamma(t)$ is small the first term dominates and the landscape resembles $f(x)$. Another way is to imagine that high values of $\Gamma(t)$ allow the computation to escape local minima by "tunneling through" hills instead of climbing over them incrementally, as SA does. This allows the algorithm to move faster and further across the landscape early in the process.

The resemblance of (3.4) to an AQC algorithm (see Chapter 2) is obvious. The problem Hamiltonians serve the same purpose. The orthogonal Hamiltonian $\mathcal{H}_D$ of QA introduces disorder to allow the heuristic search to escape local minima; this is analogous to the initial Hamiltonian $\mathcal{H}_I$ of AQC introduced to ensure that initial superposition states are equiprobable. The scaling parameter $\Gamma(t)$ creates a specific type of adiabatic path.

In this respect, QA algorithms are types of AQC algorithms. Some differences exist in the standard approaches to analysis in these two research domains, however. In AQC, one derives a bound on the probability of finishing in ground state assuming the system starts in ground state. In QA, one analyzes the probability of converging to a solution within ϵ of optimal when starting from an arbitrary state.

In fact, if the algorithm is implemented on a quantum platform, it is not in a single state at a given time t, but rather in a superposition of states, with probabilities determined by $\mathcal{H}(t)$. Unlike SA, implementations of QA on classical platforms do not typically represent a single state moving across the landscape, but instead try to capture this superposition property by representing many states in parallel.

3.2 CLASSICAL IMPLEMENTATIONS OF QA

There are many possible ways to implement a classical QA algorithm that finds the ground state of a time-dependent Hamiltonian such as (3.4). See Battaglia et al. [8], Battaglia et al. [9], Das and Chakrabarti [27], Martonak et al. [66], or Matsuda et al. [67] for examples. One common strategy is simply to carry out an exhaustive numerical calculation of the state space probabilities evolving over time. This approach is of course computationally very expensive. Gaitan and Clark [41] mention that their numerical simulations of an algorithm to find Ramsey numbers were only practical up to a limit of $n = 22$; and Farhi et al. [36] remark that their study of an algorithm for Exact Cover via numerical integration took a few months of computation time on problem sizes up to $n = 20$.

Another approach is to employ a random sampling strategy to estimate the state space probabilities at discretized times t. For example, Battaglia et al. [8] describe an approach to simulating QA via *Path Integral Monte Carlo* (PMIC), which is a type of Monte Carlo sampling method. Assuming a solution is represented by a string of n bits, this approach starts with P so-called *Trotter replicas* of the state space. The standard heuristic search approach is modified so that the main loop contains an inner loop that performs some number of Monte Carlo "sweeps" through the bits in each solution state, updating replicas by flipping bits according to the probability in (3.1). Rather than computing the bit probabilities independently, it is common practice to use a "global" cost for all P replicas, and to flip their corresponding bits simultaneously using word parallelism.

The requirement to model quantum behavior creates a high cost overhead for classically implemented QA algorithms. When considered strictly as an optimization metaheuristic, the quantum annealing approach does not appear to be competitive with standard techniques such as simulated annealing, tabu search, GRASP, integer/linear programming, and so forth.

However, QA does have some intriguing and useful properties. For example, Morita and Nishimori [74] have shown a convergence bound for QA analogous to (3.3) for SA. Assuming that $\Gamma(t)$ decreases according to

$$\Gamma(i) \geq \frac{b}{(i+1)^{c/n}} \tag{3.5}$$

for constants b, c, the algorithm guarantees to reach ground state with a probability approaching 1 as $i \to \infty$. This convergence rate is better (in n) than that of SA, although both bounds are quite pessimistic. In some restricted cases QA can be proven to converge in polynomial time when the corresponding SA approach has an exponential lower bound (see Morita and Nishimori [74]). Ohzeki and Nishimori [81] also point out that in nearly all cases where QA and SA have been compared empirically by numerical simulation or sampling, QA has shown more rapid convergence to ground state than SA. On the other hand, Battaglia et al. [8] compare implementations of SA and QA for a single large random 3SAT problem of $n = 10^4$ variables. Their version of SA outperforms QA, in contrast to the typical outcome.

Some work has been carried out to learn what properties of the landscape might drive convergence rates. Battaglia et al. [8] and Battaglia et al. [9] consider a simple landscape with two valleys separated by one high hill. They observe that their implementations of SA and QA have quite distinct performance profiles. Roughly speaking, SA performance degrades according to the *height* of barriers in the solution landscape, because there is lower probability of making a sequence of uphill moves sufficient to get over the hill. Performance of QA degrades according to the *width* of the barriers—that is, the number of steps needed to tunnel through a hill to the other side. They also look at landscapes for simple random problems and note that SA tends to follow narrow crevasses, while QA prefers to linger in middle-altitude plateaus. Slow convergence of QA in some cases may be associated with landscapes having many shallow local minima that distract the algorithm from finding lower regions to explore.

Of course there remains the tantalizing possibility that implementations of QA on quantum platforms will yield important speedups in practice over classical approaches. See Chapter 5 for some early hints about the status of this question.

The remainder of this section surveys quantum annealing algorithms that have been formulated for some well-known NP-hard optimization problems. The focus here is simply on deriving problem Hamiltonians to match the given objective functions. Little is know about good choices for transverse field Hamiltonians or cooling schedules for $\Gamma(t)$, and simple defaults are typically used in empirical work on these algorithms. Performance results from a few empirical studies are also briefly reviewed.

3.2.1 ISING MODEL AND RELATED PROBLEMS

The Ising Model is a mathematical model used for studying phase transitions and other properties of physical systems that evolve in time. This is a generalization of the native problem that D-Wave's quantum annealing chips are designed to solve; the particular version of the problem and the QA algorithm realized by those systems are discussed in Chapter 4.

The problem concerns a collection of n particles arranged on the vertices of a graph $G = (V, E)$, which is often assumed to be a d-dimensional grid. Each particle can be in one of two states, called *spins*, represented by ± 1. A *spin configuration* $s = s_1 \ldots s_n$ is an assignment of spin values to particles.

The motivating application in statistical mechanics is to characterize various properties of spin configurations, given certain *external forces* h_i applied to individual particles, and *interaction forces* J_{ij} between grid neighbors. (In cases where i and j are not adjacent we assume that $J_{ij} = 0$.) The energy of a given configuration is defined by

$$H(s) \quad = \quad \sum_i h_i s_i + \sum_{i<j} J_{ij} s_i s_j. \tag{3.6}$$

$$H(-1,-1) = -(-.5 + .3) + .4 = .6$$

$$H(-1,+1) = -(-.5 - .3) - .4 = .4$$

$$H(+1,-1) = -(+.5 + .3) - .4 = -1.2$$

$$H(+1,+1) = -(+.5 - .3) + .4 = .2$$

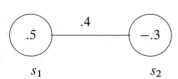

Figure 3.4: An Ising Spin Model with two variables.

Notice how the signs of weights influence the corresponding spin values: a term with negative h_i is minimized when $s_i = +1$ and a term with positive h_i is minimized when $s_i = -1$; a term with positive J_{ij} is minimized when $s_i \neq s_j$ and a term with negative J_{ij} is minimized when $s_i = s_j$. Physicists call a force such as $J_{ij} > 0$, for "different-valued" neighbors, an *antiferromagnetic coupling*; the force $J_{ij} < 0$ is a *ferromagnetic coupling*.

Several research questions in statistical mechanics require as a subproblem to find the ground state of such a system—that is, a spin configuration that minimizes $H(s)$. Istrail [53] showed that this problem is NP-hard when G is nonplanar, which includes grids of dimension three and higher. Fu and Anderson [39] have shown that the problem is NP-hard when G is a 2D grid and linear terms (nonzero h's). Olvieira and Terhal [80] show that approximation within $1/poly(n)$ accuracy is QMA-complete. Bansal et al. [7] have shown that the Ising Model has a PTAS (see Section 2.4) for planar graphs under a classical model of computation. Saket [93] describes a PTAS for Chimera graphs, which are incorporated in D-Wave platforms. (See Section 2.3 for definitions of these classes. See Section 4.1 for a description of a Chimera graph.)

Figure 3.4 presents an example instance comprising a "grid" of size $n = 2$. The four possible solutions and their costs are shown: the optimal solution is $s = (+1, -1)$.

Problem Hamiltonian.   Assume we have available a collection of n qubits $Q = q_1 \ldots q_n$ arranged on the nodes of a d-dimensional grid $G = (V, E)$, where each q_i is capable of interacting with its grid neighbors. The superposition state of Q is described by $|\Phi\rangle = |\phi_1 \ldots \phi_n\rangle$. Following the development in Section 2.2, the problem Hamiltonian for our QA algorithm is

$$\mathcal{H}_I = \sum_{i \in V} h_i \sigma_i^z + \sum_{(i,j) \in E} J_{ij} \sigma_i^z \sigma_j^z, \tag{3.7}$$

where the notation σ_i^z means that the Pauli-z operator is applied to the single qubit $|\ldots \phi_i \ldots\rangle$.

To see that this solves the problem defined by (3.6), recall that the Pauli-z operator has eigenvalues and eigenvectors defined by

$$\sigma^z |0\rangle = +1 |0\rangle = (-1)^0 |0\rangle$$
$$\sigma^z |1\rangle = -1 |1\rangle = (-1)^1 |1\rangle \tag{3.8}$$

This sets up a correspondence between $|\phi_i\rangle$ and s_i so that

$$\mathcal{H}_I |\Phi\rangle = \left(\sum_{i \in V} h_i \sigma_i^z + \sum_{(i,j) \in E} J_{ij} \sigma_i^z \sigma_j^z \right) |\Phi\rangle \tag{3.9}$$

yields the energy function

$$\mathcal{E}_I = \sum_{i \in V} h_i (-1)^{\phi_i} + \sum_{(i,j) \in E} J_{ij} (-1)^{\phi_i} (-1)^{\phi_j}$$

$$= \sum_{i \in V} h_i s_i + \sum_{(i,j) \in E} J_{ij} s_i s_j \tag{3.10}$$

Classical implementations of QA algorithms for the Ising Model have been studied empirically. Kadowaki and Nishimori [56] use numerical calculations to compare QA and SA on varying graph topologies of size up to $n = 8$, with a variety of annealing schedules for temperature T and tunneling parameter Γ. On the problems they tested, QA converged more quickly than SA. Battaglia et al. [9] describe experiments using their Path Integral Monte Carlo (PIMC) simulation of QA to compare to SA on 2-dimensional grids (giving a polynomial-time problem) of size up to $n = 80 \times 80$. They also observe that QA converges more quickly than SA on the problems they studied. Sen and Das [98] and others have observed that both QA and SA perform poorly on a one-dimensional grid model that consists of a simple chain of nodes.

Relationship to QUBO and WM2SAT   The Ising Model, defined on spins $s_i \in \pm 1$, is closely related to two problems on binary variables $x_i \in \{0, 1\}$. These latter formulations are sometimes more convenient to work with, partly because heuristic algorithms are typically implemented to return binary-valued solutions.

The first is Quadratic Unconstrained Binary Optimization (QUBO), defined as follows. Given an $n \times n$ matrix of weights Q_{ij} the problem is to assign binary values to n variables x_i for $i \in 1 \ldots n$, to minimize the function

$$f_{qubo}(x) = \sum_{i,j} Q_{ij} x_i x_j. \tag{3.11}$$

The second problem is Weighted Maximum 2-Satisfiability (WM2-SAT): You are given a boolean formula in conjunctive normal form with two variables per clause, and m clauses, where each clause has a weight w_i. The function is defined on n boolean variables $x_i \in \{0, 1\}$ where 0 and 1 correspond to False and True, respectively:

$$B(x) = w_1 C_1 \wedge \ldots \wedge w_m C_m$$
$$C_i = (\ell_{i1} \vee \ell_{i2}) \tag{3.12}$$

Given inputs $B(x)$ and T the problem is to determine whether there is an assignment of variables for which clauses of total weight at least T can be satisfied. Note that although the 2-Satisfiability problem is in P, the Maximum 2-Satisfiability problem is NP-hard. The proof is by transformation from 3CNF-SAT, with $T = 7/10$ when clause weights are equal to 1 (Garey et al. [40]).

Transformations among these three problems are straightforward using the arithmetical trick $s_i = 1 - 2x_i$ and adjusting weights accordingly. The reductions are left for the reader; or see Bain et al. [5]. Tavares [103] presents several transformations from well-known NP-complete problems to QUBO.

3.2.2   GENERAL SATISFIABILITY

The Satisfiability problem is well known: given a boolean formula $B(x)$ defined on n variables $x = x_1 \ldots x_n$, find an assignment of truth values to x such that $B(x)$ evaluates to true. We say that such an assignment *satisfies* $B(x)$.

$B(x)$ is commonly expressed in 3-variable conjunctive normal form (3CNF), which is a conjunction of m clauses, $B(x) = \bigwedge_{c=1}^{m} B_c$. Each clause contains a disjunction of three variables $B_c = (v_{c1} \vee v_{c2} \vee v_{c3})$ such that $v_{ci} = x_j$ or $\overline{x}_j$ for some $j \in 1 \ldots n$.

Here we consider more general formulations. Let G3-SAT denote a general satisfiability problem that is formulated as a disjunction of m clauses on n variables, where each clause may contain an arbitrary boolean function on three variables.

Problem formulation   To find a suitable Hamiltonian for a given G3-SAT formula, we map each boolean clause $B_c(v_{c1}, v_{c2}, v_{c3})$ to an energy function, an arithmetical expression $f_c(s_1, s_2, s_3)$ defined on spin values ± 1. The mapping should preserve the property that the clause is satisfied exactly when the function is minimized. Furthermore, for applications that require counting the number of satisfied clauses in $B(x)$, we require that the energy functions be two-valued, equal to, say, 1 when the clause is unsatisfied and 0 when the clause is satisfied, so that clauses have unit cost.

For example, recall from Section 2.3 that Exact 3-Cover (EC) is a G3-SAT problem where each clause is satisfied when exactly one of its variables is 1. (This is sometimes called 1-in-3 SAT.) This can be expressed logically as

$$B_c(v_{c1}, v_{c2}, v_{c3}) \quad = \quad (\bar{v}_{c1} \wedge \bar{v}_{c2} \wedge v_{c3}) \vee (\bar{v}_{c1} \wedge v_{c2} \wedge \bar{v}_{c3}) \vee (v_{c1} \wedge \bar{v}_{c2} \wedge \bar{v}_{c3}). \quad (3.13)$$

One way to express this with an energy function is

$$f_c(s_1, s_2, s_3) \quad = \quad (s_1 + s_2 + s_3 - 1)^2, \quad (3.14)$$

which is minimized at 0 when exactly one variable is equal to -1. If $B(x)$ consists entirely of clauses in this form, a satisfying assignment could be found using a QA algorithm with the problem Hamiltonian

$$\mathcal{H}_P \quad = \quad \sum_{c=1}^{m} (\sigma_{c1}^z + \sigma_{c2}^z + \sigma_{c3}^z - 1)^2. \quad (3.15)$$

However note that f_c is not two-valued: its range is $[16, 4, 0]$. Thus this Hamiltonian is not suitable for counting unsatisfied clauses when $B(x)$ is unsatisfiable. Hen and Young [50] define an alternative energy function with the necessary property:

$$f_c(s_1, s_2, s_3) = (5 - s_1 - s_2 - s_3 + s_1 s_2 + s_2 s_3 + s_1 s_3 + 3 s_1 s_2 s_3) / 8. \tag{3.16}$$

This function evaluates to 0 when one term equals 1, and evaluates to 1 otherwise.

There are 256 possible boolean functions on three variables: with a little thought it is not difficult to construct suitable energy functions as needed. It is convenient to represent truth values as binary values $x_i \in \{1, 0\}$ (which can be used with both boolean and arithmetic operators) and then map to spins $s_i \in \{-1, +1\}$ via $x = (1 - s)/2$. Here are some examples.

- Bain et al. [5] show how functions on spins $s = \pm 1$ can be used to represent basic logical operators as in Table 3.1.

Table 3.1: Representing logical operators with spin functions. The function is minimized at 0 exactly when the boolean expression is true. (From Bain et al. [5], Table 1)

Logical operator	Ising spin function
$z = \neg y$	$1 + s_z s_y$
$z = y_1 \wedge y_2$	$3 - (s_{y_1} + s_{y_2}) + 2 s_z + s_{y_1} s_{y_2} - 2(s_{y_1} + s_{y_2}) s_z$
$z = y_1 \vee y_2$	$3 + s_{y_1} + s_{y_2} - 2 s_z + s_{y_1} s_{y_2} - 2(s_{y_1} + s_{y_2}) s_z$

- $\text{ODD}(x_1, x_2, x_3)$ is satisfied when the bits have odd parity. The function

$$f_{odd}(s_1, s_2, s_3) = s_1 \cdot s_2 \cdot s_3 \tag{3.17}$$

is equal to +1 for even parity and to -1 for odd parity.

- $\text{NOTALLEQUAL}(x_1, x_2, x_3)$ is satisfied when one or two bits equal 1, but *not* satisfied by 000 or 111. The function

$$f_{nae}(s_1, s_2, s_3) = s_1 s_2 + s_2 s_3 + s_1 s_3 \tag{3.18}$$

evaluates to -1 when the boolean expression is satisfied and to 3 otherwise.

- The expression $((b_1 \wedge b_2) \vee b_3)$ can be mapped to

$$f(s_1, s_2, s_3) = \frac{s_1 + s_2 - s_1 s_2}{4} + \frac{s_3 + s_1 s_2 s_3}{2}, \tag{3.19}$$

which evaluates to 0 when the expression is satisfied and to 1 otherwise.

Theoretical and empirical results concerning QA algorithms for solving Satisfiability problems are reviewed in Section 2.3.1. (Although their authors refer to them as AQC algorithms, they qualify as QA algorithms under the criteria outlined at the beginning of this chapter.)

3.2.3   TRAVELING SALESMAN PROBLEM

Given a graph $G = (V, E)$ with weights w_{ij} on edges, the Traveling Salesman Problem is to find a Hamiltonian cycle of G with of minimum total edge weight. (Here we consider the symmetric version of the problem, defined in undirected graphs.)

Battaglia et al. [8] describe simulation experiments to compare SA to an implementation of QA using their PIMC approach described earlier. To formulate the problem they represent G with a weighted adjacency matrix D_{ij} and represent a tour T using an adjacency matrix $U_{ij} \in 0, 1$. The problem is to find an assignment of binary values to T to minimize the objective function

$$C(t) = \frac{1}{2} \sum_{i<j} d_{ij} u_{ij}. \qquad (3.20)$$

This objective function can be converted into a problem Hamiltonian in a straightforward way. The authors compare their implementation to a simulated annealing solver on a single instance of size $n = 1002$, for which the optimal solution is known. Therefore they are able to look at the distance from optimal solution as a function of loop iterations; they observe that QA converges to optimality faster than SA.

Note that this is an example of a *constrained* optimization problem, since not every assignment of bits to T represents a valid tour of G. The authors handle this problem by restricting their simulation code to start with a valid tour and to use a particular neighborhood rule (called a two-opt move) that only permits steps to valid tours.

The usual technique for expressing a constrained problem in an unconstrained framework is to introduce *penalty terms* that associate infeasible solutions with high objective function costs. It is not clear in general how to incorporate combinatorial constraints into the Hamiltonians that describe AQC and QA algorithms. This question has received little attention in the AQC and QA literature.

3.2.4   FACTORING INTEGERS

The Factoring problem is, given an n-bit integer N, to report at least one of its nontrivial prime factors. This problem is not known to be NP-complete; nor has it been shown to be in P. This problem is of great practical and popular interest because the security of the current version of the RSA encryption algorithm rests on the presumed difficulty of factoring large integers.

Xu et al. [109] describe a quantum annealing algorithm for factoring that is a simplification of an earlier proposal by Schaller and Schützhold [96]. They (Xu et al. [109]) implemented their algorithm on a quantum annealing processor built using liquid crystal NMR technology (different from D-Wave's approach), described further in Section 4.4. Their 4-qubit processor successfully factored the number 143 into 11 and 13.

To cast factoring as an optimization problem we start with the objective function $f_N(P, Q) = (N - PQ)^2$. That is, for given N, the problem is to find positive integers P and Q to minimize this cost. Suppose that N has d bits and P and Q are of length $d/2$. We can write out the multiplication table for the individual bits, including carry bits $Z = z_{ij}$ as follows (reproduced from Xu et al. [109] for the specific value $N = 143$).

				1	p_2	p_1	1
				1	q_2	q_1	1
				1	p_2	p_1	1
			q_1	$p_2 q_1$	$p_1 q_1$	q_1	
		q_2	$p_2 q_2$	$p_1 q_2$	q_2		
	1	p_2	p_1	1			
z_{67}	z_{56}	z_{45}	z_{34}	z_{23}	z_{12}		
z_{57}	z_{46}	z_{35}	z_{24}				
1	0	0	0	1	1	1	1

Note that certain entries in the table can be inferred from the problem statement; for example N, P, and Q must be odd, so the right-most bits are 1. Xu et al. [109] construct bit-wise final Hamiltonians $\mathcal{H}_i$ from columns in this table, using algebraic tricks to keep the number of pairwise interactions low. Transforming from binary to spin values $p_i = (1 - \hat{p}_i)/2$, they obtain the objective function

$$
\begin{aligned}
\mathcal{F}_{143}(P, Q) \;=\; & 5 - 3\hat{p}_1 - \hat{p}_2 - \hat{q}_1 + 2\hat{p}_1\hat{q}_1 - 3\hat{p}_2\hat{q}_1 \\
& + 2\hat{p}_1\hat{p}_2\hat{q}_1 - 3\hat{q}_2 + \hat{p}_1\hat{q}_2 + 2\hat{p}_2\hat{q}_2 + 2\hat{p}_2\hat{q}_1\hat{q}_2.
\end{aligned} \tag{3.21}
$$

They solve the corresponding Hamiltonian column-by-column starting from low order bits, and verify that the computation is correct at each iteration. This adds $O(n^3)$ classical overhead to computation time.

The main obstacle to generalizing this achievement to obtain a usable tool for taking down the Internet is the difficulty of casting the problem into suitable local form. Xu et al. [109] are able to find a formulation of the problem for $N = 143$ with terms containing at most three variables, but in general there is no known systematic method for converting an arbitrary integer N to an objective function with small enough locality to be represented by a Hamiltonian that is realizable in practice.

CHAPTER 4

The D-Wave Platform

Previous chapters have set out the theoretical framework that underlies the adiabatic model of computation and the quantum annealing paradigm. The remainder of this book considers practical issues: this chapter gives an overview of D-Wave[1] quantum annealing systems, and Chapter 5 surveys experimental research to study their properties.

Section 4.1 presents a walk-through of the user experience when solving an optimization problem. Section 4.2 describes the D-Wave technology stack. Because of space limitations it is only possible to give a bare sketch of the general properties of these components: for more technical details see Bunyk et al. [19], Harris et al. [48], Harris et al. [49], and Johnson et al. [55]. Section 4.3 discusses challenges to success when using this approach to quantum computation. The chapter concludes with a brief survey of alternative quantum annealing systems built by other research groups. The discussion throughout is aimed at a non-physicist and non-specialist in quantum computing.

4.1 THE USER'S VIEW

A D-Wave platform comprises two main components:

1. A processor chip ("the hardware") that solves Ising Model (IM) problem instances by physical realization of a quantum annealing algorithm. The chip is mounted in a dilution refrigerator and supercooled to a target operating temperature below 20mK, which is necessary to achieve quantum effects. The chip subsystem is housed within many layers of shielding to protect against various types of environmental noise.

2. A conventional (Intel) front end server connected to the chip via control lines and an I/O subsystem. The front end receives instructions from the user and is accessed using a cloud computing model that supports job queuing and scheduling. The front end sends a problem instance to the hardware and receives a set of solutions in reply.

We start with a walk-through of the user experience to solve a combinatorial optimization problem, sketched in Figure 4.1. Note that quantum annealing can be used in many other domains for which optimization is a subproblem. Chapter 5 surveys the wider field of applications.

Suppose the user has an instance I in hand for a given problem P. If I is in *native* form, it can be submitted to the front end to be loaded onto the quantum annealing chip for direct solution in hardware. The hardware can be programmed via a low-level *Quantum Machine Instruction*

---

[1]D-Wave, D-Wave One, D-Wave Two, Vesuvius, and QSage are trademarks of D-Wave Systems Inc.

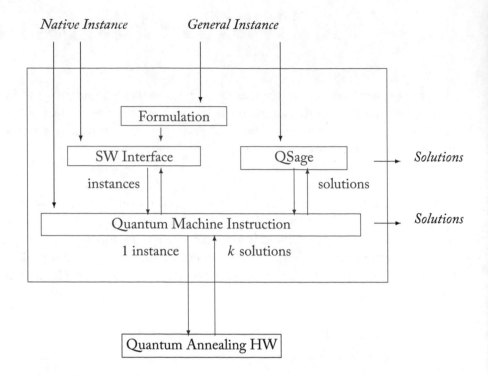

Figure 4.1: An overview of the D-Wave Quantum Annealing System. Several paths from input instance to solution can be identified.

interface, or through a *Software Interface* containing, for example, C, Python, and Matlab programming tools. If the instance is not in native form, the user has two options: translate it into a native instance (software tools to help with this task are part of the *Formulation* box in Figure 4.1); or submit it as is to a hybrid classical-quantum solver called QSage. We consider each option in turn.

Native instances.   Recall the Ising Model (IM) from Section 3.2.1: Given a set of real weights h_i and J_{ij}, find an assignment to n spin variables $S = s_1 \ldots s_n$, with $s_i \in \{\pm 1\}$, to minimize the objective function

$$f_{im}(S) \;\; = \;\; \sum_i h_i s_i + \sum_{i<j} J_{ij} s_i s_j. \tag{4.1}$$

This problem is directly implemented in the quantum annealing hardware: that is, the weights h_i and J_{ij} correspond to electronic signals called *biases* that are applied to the qubits and to the couplings that connect pairs of qubits. The qubits work as quantum particles in a quantum annealing process to find a low-energy state that corresponds to a low-cost solution to the objective function.

Some restrictions to the general problem are imposed by the hardware design. First, the h_i and J_{ij} weights must match the physical connectivity structure on the chip. Qubit connectivity is based on a *Chimera* structure denoted C_k. As shown on the right side of Figure 4.5, each cell in a Chimera structure is a complete bipartite graph $K_{4,4}$ on 8 nodes. Cells are connected in a $k \times k$ grid, with four qubits connected to horizontal neighbors and four qubits connected to vertical neighbors. D-Wave One chips are based on a C_4 graph with 128 qubits, and D-Wave Two chips are based on C_8 with 512 qubits. Figure 4.6 shows an optical image of a D-Wave Two chip where the 8×8 grid of cells is visible. Chimera graphs are non-planar; for example a C_k contains a 3D grid of size $k \times k \times 4$ as a subgraph. The treewidth of a C_k graph (a graph property that is related to problem complexity) is $4k$. Internal nodes have degree 6 and boundary nodes have degrees 5 and 4.

The current chip fabrication process leaves some number of qubits and couplers inoperable. Therefore each individual chip has a specific *hardware working graph* $H \subset C_k$. The weights h_i and J_{ij} in (4.1) must be set to 0 if the corresponding qubits or couplers are not present or nonfunctional in the working graph. An example H based on C_8 is shown on the right side of Figure 4.2.

Another restriction concerns the precision and range of weights h_i and J_{ij}. Although the front end accepts arbitrary floating point values as weights, the analog control circuitry that applies the biases to qubits and couplers has fixed range and limited precision. Therefore input weights may not be mapped with sufficient fidelity to the hardware, causing it to solve a different problem than the one specified. The limits on viable weight sets depend on properties of the input and are difficult to characterize. More about this in Section 4.3.

Mapping a general instance to native form.   Transformation of a general optimization problem instance to a native instance is a two-step process: first, translate the problem to the Ising Model (IM), and second, map the general IM connectivity graph to the hardware working graph.

Conceptually it is not difficult to extend the standard transformation techniques of NP-completeness theory to work for optimization problems. Thus an instance A for any given problem can be mapped to an instance M for IM so that an optimal solution to M yields an optimal solution to A, with no more than polynomial expansion in problem size. As discussed in Section 3.2.1, IM is closely related to Quadratic Unconstrained Boolean Optimization (QUBO) and Weighted Maximum 2-Satisfiability (WM2SAT), which are defined on binary variables in $\{0, 1\}$ instead of on spins ± 1. Transformation of A to one of these problems may be more convenient in some cases. Boros et al. [17] and Tavares [103] give many example transformations from NP-hard problems to QUBO. Because of limited chip sizes we are interested in compact transformations that yield low expansions in problem size; it appears that very little is known about optimizing transformations in this way.

The second step is to map the instance onto the hardware. Let $G = (V_g, E_g)$ denote the connectivity structure of a general IM problem, where vertices corresponds to problem variables s_i and edges $E_g = (v_i, v_j)$ exists wherever $J_{ij} \neq 0$. Let $H = (V_h, E_h)$ represent the hardware working graph, a subgraph of C_k.

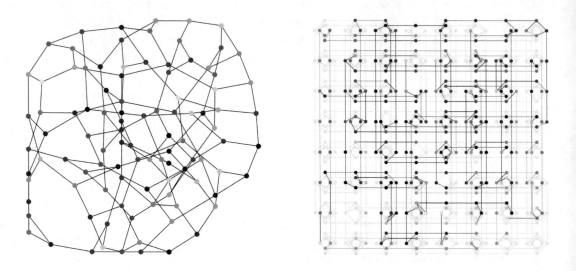

Figure 4.2: Minor-embedding a general graph G into a hardware working graph H. Each source vertex is mapped to chain of like-colored vertices (Image source: Aidan Roy).

The problem is to find a *minor-embedding* of G onto H, as shown in Figure 4.2. A minor of H is any graph H' that can be constructed from H by collapsing two adjacent vertices into one (combining their incident edges), or by removing an edge. Equivalently, we can construct a minor-embedding of G onto H as follows:

1. Source vertex $v_i \in V_g$ can be split into multiple vertices to form a connected subgraph T_i such that $T_i \subseteq H$. We say the edges of T_i form a *chain* that must be mapped onto H. (Although the term "chains" implies a linear structure, arbitrary connectivity is allowed).

2. Source edge $(v_i, v_j) \in E_g$ must map to an edge $(w_i, w_j) \in H$ such that $w_i \in T_i$ and $w_j \in T_j$. These are *problem edges* that map directly to physical edges of H

3. Weights on chain edges are set to large-magnitude negative values (ferromagnetic couplings) so that all vertices in T_i have the same spin in the optimal solution.

4. Weights on problem edges are as defined in the original problem; weight h_i may be assigned to one vertex in T_i or split among multiple vertices.

Choi [24] gives a polynomial algorithm for minor-embedding a complete graph of n vertices onto the upper diagonal of a $k \times k$ Chimera graph with $k = 4n$. Thus a C_8 can hold a complete graph on 32 vertices. Choi [25] also shows that the Chimera graph is optimal for this purpose, over all graphs that meet certain sparsity and locality conditions imposed by engineering constraints on the hardware. The complexity of finding an optimal minor-embedding from an arbitrary graph to a Chimera graph is unknown (note this is not exactly analogous to a graph isomorphism problem because the destination C_k is fixed).

While Choi's construction is optimal for minor-embedding in C_k, it is not clear how to adapt the algorithm to a given $H \subset C_k$. Also this approach tends to create over-large chains in practice, which can be more difficult to solve to optimality. The problem is related to limitations on biases: if chain weights are too small, the spins assigned to vertices of T_i may disagree in the optimal solution, which creates an infeasible result. But too-large chain weights "compress" the range of problem weights when scaled, which may create precision errors. In practice the user may employ a software tool that uses a heuristic search approach to find good minor-embeddings in a given hardware graph. Some pre- and post-processing strategies are also available for mitigating problems related to chains.

QSage hybrid optimizer.   Instead of being translated to native form, a general problem instance can be submitted to the QSage solver. This hybrid software/hardware tool accepts the instance plus an objective function $f(x)$ defined on n binary variables, where n must be no more than the number of qubits in the working graph H. The function evaluation code is used as an oracle, which means that the solver does not rely on domain-specific assumptions about the problem.

This solver uses a tabu search approach—a type of heuristic search—to walk through the solution landscape looking for a global minimum. At each iteration i with current solution x_i, tabu search generates a list N of n neighbor solutions. It checks those neighbors against a tabu list T of recent moves in the state space, and strikes off from N any that have been considered in the most recent m iterations. The node with smallest cost among those remaining in N is then accepted as the next solution x_{i+1}. This use of a tabu list to avoid repeat visits to recently considered solutions forces the solver to move out of local minima by accepting an uphill move if that is the best option in N.

The QSage optimizer incorporates an additional feature to this approach by querying the hardware at each iteration. Given the solutions in N and their objective function values, it constructs a native instance for the hardware, which may be considered an approximation of the state space topology in that region. The hardware returns a sample N' of k additional solutions, and the combined set of solutions $N'' = N \cup N'$ is checked against T. The idea is to accelerate convergence by using hardware queries to jump to low-lying regions of the solution space for further exploration.

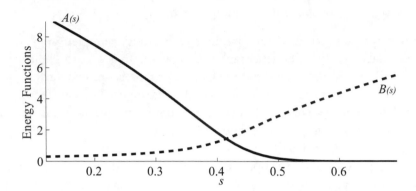

Figure 4.3: The annealing path is described by a pair of functions $A(s)$ and $B(s)$ where $s(t) : 0 \to 1$ as $t : 0 \to t_f$ for some elapsed time t_f.

Annealing in hardware.   Once a native instance is formulated by one of the above methods, it can be sent to the quantum annealing hardware for solution. The quantum annealing algorithm used by D-Wave incorporates the initial Hamiltonian

$$\mathcal{H}_I \quad = \quad \sum_i \sigma_i^x, \tag{4.2}$$

and the problem Hamiltonian

$$\mathcal{H}_{IM} \quad = \quad \sum_i h_i \sigma_i^z + \sum_{i<j} J_{ij} \sigma_i^z \sigma_j^z, \tag{4.3}$$

where h_i and J_{ij} are constrained to match the hardware working graph. The annealing path is defined by a pair of *envelope functions* $A(s)$, $B(s)$ shown in Figure 4.3, where $s(t) : 0 \to 1$ as $t : 0 \to t_f$ for a total transition time t_f. The rate-of-change function $s(t)$ is not exactly linear in t but instead "slows down" somewhat in the middle of the transition: see Harris et al. [49] for technical discussion.

Thus D-Wave quantum annealers implement the following Hamiltonian:

$$\mathcal{H}(s) \quad = \quad A(s)\mathcal{H}_I + B(s)\mathcal{H}_{IM}. \tag{4.4}$$

The computation takes place in stages as follows. Table 4.1 shows some representative times for three chips.

1. **Programming/Initialization.** The weights h_i, J_{ij} are loaded onto qubit and coupler biases. The qubits are placed in superposition according to $\mathcal{H}_I$ which is scaled by $A(0)$. Programming raises the chip temperature, so this step also includes a wait time for the chip to cool back down to operating temperature. As shown in Table 4.1, initialization dominates the computation time but this cost has decreased with succeeding chip models.

2. **Anneal.** A transition takes place whereby the forces on the qubits change according to the path functions in Figure 4.3. Anneal time t_f can be set by the user: Table 4.1 shows minimum possible settings for V5 and V6.

3. **Readout.** The transition ends. The qubits now have classical spin states according to $\mathcal{H}_{IM}$, which is scaled by $B(1)$. Qubit values are read to yield the solution S. This step also includes a short wait time for the chip to cool down for resampling.

4. **Resampling.** Since the chip operates in an open system, there is always a positive (sometimes significant) probability that the computation does not finish in ground state. Given the relatively high initialization times it is cost-effective to repeat the anneal-readout cycle many times, typically $k = 1{,}000$ or $k = 10{,}000$. Thus a given chip requires time $T_k = T_p + k(T_a + T_r)$ to return a sample of k solutions to one instance. The last column in Table 4.1 shows total time to sample $k = 1{,}000$ solutions.

Table 4.1: Computation times for a Model One C_4, and two Model Two C_8 chips in the Vesuvius series. The second column shows the number of working qubits on the chip. The next three columns show times for initialization, anneal, and readout. The last column shows total time to sample $k = 1{,}000$ solutions for one instance (Sources: Bain et al. [5] (One), McGeoch and Wang [70] (Two)).

Chip	Qubits	T_p	T_a	T_r	T_{1000}
One	102	270ms	1ms	1.5ms	2770ms
Two V5	439	201ms	$20\mu s$	.29ms	491ms
Two V6	502	36ms	$20\mu s$	.13ms	196ms

4.2   THE TECHNOLOGY STACK

This section gives a quick overview of the D-Wave technology stack. Three layers are highlighted here:

- **Qubits.** Section 4.2.1 describes the superconducting flux qubits used to represent quantum state.

- **Connection topology.** Section 4.2.2 discusses features of the Chimera structure.

- **Control systems.** Section 4.2.3 looks at systems for reading, writing, and manipulating qubit states.

4.2.1 QUBITS AND COUPLERS

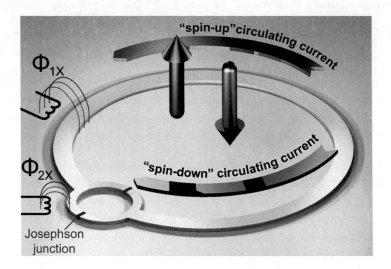

Figure 4.4: Each qubit is a loop of niobium controlled by Josephson junctions and biases represented by Φ's. The direction of current in the loop corresponds to the state of the qubit. Niobium loops can be place in superposition (current running in both directions) when supercooled to temperatures below 20 mK (Image source: Mark Johnson).

A D-Wave chip holds an arrangement of *superconducting flux qubits*, specifically CCJJ rf-SQUIDs.[2] As shown in the simplified diagram of Figure 4.4, each qubit is made of niobium (Nb) in the shape of a double ring interrupted and controlled by two Josephson junctions. The flux (flow of current) in the ring depends on biases represented by Φ values. For example in this diagram Φ_{1x} corresponds to a weight h_i and Φ_{2x} to the value of $A(s)$ in the path function. Current can flow counterclockwise, or clockwise, or in both directions at once, when the qubit is in quantum superposition. Flux creates spins $|\uparrow\rangle$ and $|\downarrow\rangle$ that represent the state of the qubit.

Couplers form the interconnections between pairs of qubits. Couplers are also fabricated from loops of niobium, but are simpler and have less control circuitry than qubits. Qubits and couplers are etched onto silicon wafer circuits consisting of four Nb layers separated by insulation.

Harris et al. [49] describe some pros and cons of choosing to fabricate with superconducting flux qubits as opposed to other options (see Ladd et al. [63] for a survey of the huge number of available qubit technologies). One drawback is that the superconductivity necessary to realize quantum superposition in niobum is only possible when the material is cryogenically frozen

---

[2]Compound-compound Josephson junction, radio frequency, superconducting quantum interference devices.

to temperatures below 20mK. This technology is highly sensitive to certain types of noise but robust against other types of noise, both external and internal (crosstalk). This tradeoff makes deployment in large scale quantum processors possible, but the chips must be operated in super-cooled refrigerators within many layers of shielding to protect from the so-called *energy bath* of the environment.

Perhaps the most important positive feature of the rf SQUIDS deployed in D-Wave chips is that they allow a high level of tunability *in situ*, which means it is possible to correct for some types of fabrication errors (inevitable under current fabrication standards). Tunability also means that qubits are runtime programmable, and that some types of fabrication anomalies can be corrected dynamically. Each qubit has six "control knobs" onboard the chip that carry signals from the processor to adjust its physical parameters. This feature produces acceptable levels of consistency and regularity during the quantum annealing process.

4.2.2 TOPOLOGY

To realize the Chimera topology in hardware, groups of eight qubits are arranged in criss-cross fashion to form the Chimera cells, as shown in Figure 4.5. Each niobium ring is stretched into a long narrow loop that crosses four other qubits. The couplers (shown in dark grey) connect pairs of qubits at their intersections. The control circuitry (bias controls and I/O) is located in the grid interstices. Figure 4.6 shows an image of the Vesuvius 6 (V6) chip, where the C_8 grid structure can be seen.

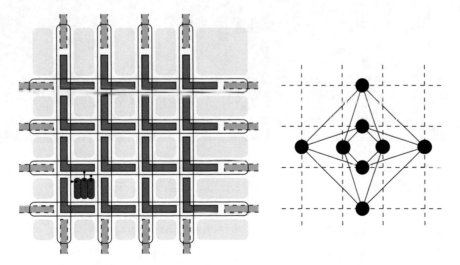

Figure 4.5: A single Chimera cell. Each cell contains a grid of four horizontal loops and four vertical loops of niobium. Couplers are shown in dark grey, and the control circuitry is in darkest grey. The qubits in each cell form a complete bipartite graph $K_{4,4}$ (Image source: Paul Bunyk).

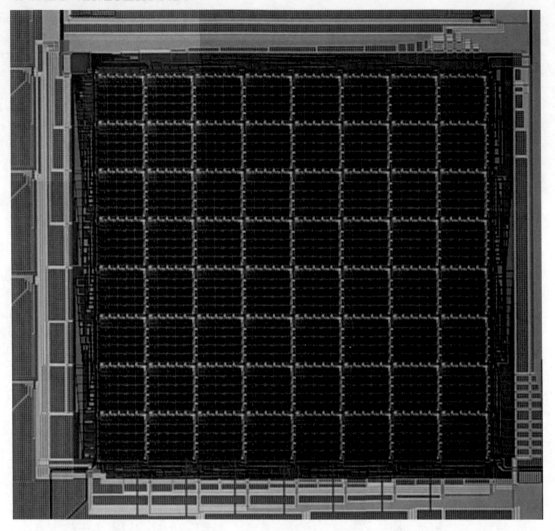

Figure 4.6: An image of the V6 chip based on a C_8 Chimera structure. Peripherals and I/O devices are arranged on the edges and in cell interstices (Image source: Paul Bunyk).

Bunyk et al. [19] list several features that make the Chimera structure attractive for its purpose, including:

- The 4×4 cell structure can be achieved within current physical limits on qubit and coupler lengths. This grid structure is also robust against certain types of crosstalk.

- The design scales well in the sense that larger C_k graphs can be built without increasing individual qubit connectivity.

- The layout can be realized by a fabrication containing at most two qubit layers.

- The native problem solved by the chip (Ising Model) remains NP-hard when restricted to this (nonplanar) structure.

- As mentioned in Section 4.1, a complete graph on k vertices can be minor-embedded into the upper diagonal of C_{k^2}, which is optimal.

For more discussion of topology and architecture considerations see Bunyk et al. [19].

4.2.3 CONTROL CIRCUITRY

As mentioned previously, each qubit in a D-Wave Two chip is connected to six "control knobs" called Φ-DACs, which carry signals from the processor and perform digital-to-analog conversion and signal latching. One Φ-DAC carries the problem bias corresponding to h_i, and the other five are used to tune and correct for variations from specification that arise in fabrication.

To load a problem onto the chip, the front end transfers weights J_{ij}, h_i into Φ-DAC biases. The processor initializes qubit states by "dialing down" the scale of $\mathcal{H}_{IM}$ and "dialing up" the scale of $\mathcal{H}_I$, according to $A(0)$ and $B(0)$. The qubits relax into their superposition ground states according to the initial Hamiltonian, essentially "ignoring" the problem Hamiltonian. The anneal process involves adjusting biases according to (4.4).

At the end of the computation it is necessary to measure the qubits and read the answer. Berkeley et al. [11] describe the qubit-addressable readout circuitry employed in D-Wave Two platforms. Each SQUID qubit is connected to a dc (direct current) SQUID, which has a voltage state that depends on the flux imparted by the qubit and which can be read by applying a small voltage pulse. Simply coupling the two SQUIDS does not work, however, because even low pulse voltages can alter the flux states of nearby qubits. The remedy described in Berkeley et al. [11] involves insertion of a third SQUID called a quantum flux parametron (QFP) that serves as a buffer and allows the dc SQUID to safely latch the qubit state.

Robustness against decoherence.   It is generally recognized in quantum computing that vulnerability to decoherence—the tendency of qubits to lose their fragile entangled states—is especially problematic at read-time, since qubits must necessarily interact with their environment while being read. However, in the quantum annealing framework implemented on D-Wave chips, the qubits finish the computation in classical states described by $\mathcal{H}_F$, which are quite robust against decoherence. At sufficiently low temperatures, and absent strong excitations, classical qubits will remain in ground state for as long as is needed.

It is also generally accepted in quantum computation that computation times must be much smaller than decoherence times to ensure adequate success probabilities. Childs et al. [21] show, however, that the adiabatic model is inherently robust in this regard: the success of the computation is not related to the phase of the ground state, and therefore success probabilities are not affected by dephasing (a type of decoherence). They also show that decoherence has a less significant effect on success probabilities at low operating temperatures such as employed in D-Wave architectures. Indeed at low temperatures decoherence tends to favor transitions toward ground state, which is the desired outcome.

Boixo et al. [14] describe experimental research that supports this theoretical argument. They observe that decoherence occurs in the energy basis, not in the computational basis (where readout occurs); this allows computation times to be much larger than decoherence times with little effect on the outcome.

4.3 CHALLENGES

Although decoherence is less problematic than in the gate model, a number of other significant physical and practical challenges do arise that reduce the probability of successful computation during operation. This section surveys some obstacles to success and some strategies for overcoming them.

Properties of the input.   The adiabatic theorem provides a lower bound on necessary transition time t_f, in terms of the minimum spectral gap δ_m that occurs during the transition. Some algorithmic strategies for "growing" the spectral gap by changing the adiabatic path, the initial Hamiltonian, or the final Hamiltonian, are discussed in Section 2.3.1.

In practice, actual computation of δ_m for a given instance is out of reach for all but a few simple scenarios. Therefore it is necessary to guess a value of t_f that would ensure successful computation in a closed system. A wrong guess increases the probability that the solution returned is not optimal.

The energy bath.   As mentioned in Chapter 3, no physical quantum computer can be perfectly isolated from its environment. Therefore it must run in an energy bath that introduces noise, resulting in lowered success probabilities compared to computation in closed systems.

The problem arises when the noise is greater than a threshold determined by the spectral gap δ_m. In such a situation the quantum particle system can't tell the difference between ground state and the first excited state, and may jump to the latter and be carried to a higher-energy excited state. State transitions (including transitions to lower-energy states) are more difficult to make later in the process, as the quantum annealing algorithm moves from "more quantum" $\mathcal{H}_I$ to "less quantum" $\mathcal{H}_{IM}$.

Interestingly, Dickson et al. [31] observe that a small amount of thermal noise can in fact help the computation succeed, by allowing the system to return to ground state later in the anneal. They compare predicted success probabilities in an adiabatic environment to measurements using a 16-qubit section of a D-Wave One chip, for a problem with an extremely small spectral gap. Success probabilities do not decrease monotonically as temperature increases: in some cases raising the temperature yields speedups (from higher success probabilities and fewer samples) by factors up to 1,000x over a theoretical closed system.

As an engineering problem the success probability may be increased by strategies such as lowering the operating temperature, increasing the amount of shielding, and re-designing the chip to reduce crosstalk within the circuit and its controllers. It is also possible to modify the shape of the adiabatic path determined by $A(s)$, $B(s)$, and $s(t)$, as a part of the system calibration procedure.

Some tactics are also available for counteracting this problem during computation. One obvious remedy against low success probability is to repeat the computation. It is standard practice to take many solution samples (typically 1,000 or 10,000) per input. Another approach is to increase anneal time t_f to improve success probabilities. On the other hand a too-slow transition creates more opportunity for errors to occur due to environmental noise. Boixo et al. [14] observe empirically that the optimal tradeoff between transition time and number of samples in current models tends to occur at minimum anneal settings.

Intrinsic control errors (ICE).   Another set of challenges arises from physical limitations of the control circuitry. Although the D-Wave front end processor accepts weights specified as single-precision floating point numbers, the digital-to-analog conversion does not support perfect fidelity in translation to biases. Environmental noise also creates random fluctuations in biases. The result is a "physical" problem Hamiltonian that does not match the "logical" Hamiltonian supplied by the processor, and that changes slightly each time an anneal is run.

The engineering challenge is to reduce ICE by any means possible. As a general rule, successive models of D-Wave chips are designed to incorporate new strategies for suppressing ICE and thereby increasing the range and precision of qubit and coupler biases. This allows a wider area of the problem space to be solved with higher success probabilities.

To circumvent ICE during computation, it is sometimes possible to boost success probabilities using simple post-processing strategies. For example given hardware solution S it is easy to check for better solutions among those within Hamming distance one of S. Boixo et al. [15] describe the application of this idea to a simple input class. Another possibility is to greedily search the solution space near S.

Pudenz et al. [83] describe a suite of error suppression techniques using redundant qubits to overcome ICE and environmental noise. These tactics can dramatically improve success probabilities, at the cost of requiring additional qubits in the problem formulation. One strategy involves allowing redundant qubits to vote on the correct answer; another involves using auxiliary qubits to increase the probability that redundant qubits will agree.

Another approach is to modify the problem Hamiltonian to avoid possible sources of systematic bias. An example is *gauge transformation* (see Boixo et al. [15]), which works as follows. Given an Ising Model instance (J_{ij}, h_i), it is possible to create a nearly identical instance by flipping the sign of a single h_i and the signs of all incident J_{ij}. To transform the solution back to the original problem it is only necessary to change the sign of s_i. (It is also possible to flip signs on J_{ij}'s with appropriate adjustments to incident vertices.) A gauge transformation involves flipping signs of random subset of problem weights, in order to obtain the same problem with different

Hamiltonian. Changing the problem in this way may increase δ_m and thereby increase success probabilities; it may also allow the computation to circumvent possible sources of systematic bias. On the other hand re-running a gauge-transformed instance incurs extra overhead in programming time; it may be more cost-effective to simply take extra samples.

On the theoretical side, Childs et al. [21] note that AQC may be insensitive to certain types of control errors. Suppose the algorithm is modified by an "error Hamiltonian" so that the actual Hamiltonian solved by the physical system is of the form $\mathcal{H}(t) + \mathcal{H}_{err}(t)$. The authors report on simulation experiments showing that under a variety of scenarios $\mathcal{H}_{err}(t)$ does not do significant harm to success probabilities in the original problem.

4.4 SOME ALTERNATIVE QUANTUM ANNEALING SYSTEMS

Other initiatives in building quantum computers in the AQC framework have been reported in the research literature. The main differences between these various approaches are in the technology adopted for physical realization of qubits and in the target problem scope: some groups are working toward universal computation and others aim to solve specialized problems distinct from those addressed by D-Wave.

An NMR-based quantum optimizer.   Steffen et al. [102] report on the first successful computation by an AQC-based computer using nuclear magnetic resonance (NMR) techniques to realize three qubits. This device can be operated at room temperature.

The device implements an AQC algorithm to solve the (NP-hard) Maximum Cut problem: Given a graph of n nodes with weighted edges, a maximum cut is a partition of the nodes into two sets that maximizes the sum of edges between the sets. The authors also introduce node weights to break solution symmetries. Given weights w_i and w_{ij}, the problem Hamiltonian is

$$\mathcal{H}_{MC} = \sum_i \frac{w_i(I - \sigma_i^z)}{2} + \sum_{i<j} w_{ij}\frac{(I - \sigma_i^z \sigma_j^z)}{2}, \tag{4.5}$$

where I is the identity matrix and σ_i^z is the Pauli-z operator applied to spin i. For $n = 3$ (the problem size in their computation) this amounts to finding the pair of edges in a triangle having maximum weight.

Their implementation in fact uses a discretized version of the algorithm that approximates the continuous Hamiltonian using a series of time steps $t = i/M$ for $i \in [0, M]$ and $M \in [15 \dots 100]$. They observe that observed performance with respect to M agrees well with theoretical models.

NMR for factoring.   Xu et al. [109] describe the use of an adiabatic quantum computer to factor the integer 143. Their algorithm is sketched in Section 3.2.4. The computer used for this computation has four qubits; each is represented of hydrogen nuclei in molecules of 1-bromo-2-chlorobenzene. Each spin is manipulated using bursts of radio waves.

CHAPTER 5

Computational Experience

This chapter surveys published research describing experiments on, and experience with, a series of D-Wave platforms. This empirical work can be roughly organized around three questions:

1. *What Problems can it Solve?* Section 5.1 surveys applications of D-Wave quantum annealing systems to a variety of problem domains.

2. *Is it quantum?* Section 5.2 describes experimental work to look for evidence of quantum properties in D-Wave hardware.

3. *How fast is it?* Section 5.3 surveys work to study hardware and system performance as compared to classical approaches.

5.1 WHAT PROBLEMS CAN IT SOLVE?

We start by considering what kinds of problems cannot be solved to optimality on current versions of D-Wave chips. One obvious limitation is problem size. Any given NP-hard problem instance can be translated to an Ising Model instance with no more than polynomial expansion in problem size, and any Ising Model instance on n variables can be minor-embedded onto a Chimera graph using $O(n^2)$ qubits in the worst case. The constant factors in those polynomials become critical to determining the largest problems that can be solved in practice. As a point of reference, the largest complete graph that can be represented in a C_8 Chimera graph containing 512 qubits is of size $n = 32$. Embeddability corresponds roughly to edge count, so larger sparse graphs can be accommodated.

It is certainly possible to devise heuristic methods that can decompose a large instance into smaller segments for direct solution in hardware, and then stitch the pieces together to obtain a (non-optimal) solution to the original problem. Examples of this approach are discussed in Sections 5.1.3 and 5.1.1. However this idea has not yet been fully explored with respect to the larger universe of combinatorial optimization problems: more work is needed.

As mentioned in Section 4.3, some input properties are known—including range and precision of weights, ground state degeneracy, and the spectral gap—that can affect the probability that any single anneal will find an optimal solution. Successive models of D-Wave chips have been designed not only for larger qubit counts, but also for more robust performance with respect to these challenges; future models are expected to show more progress along these lines. More experimental work is needed to sort out how interactions between inputs and hardware properties affect success probabilities and overall computation times.

We may also consider whole categories of application areas that *could* be tackled by D-Wave solution methods, but have not yet been fully explored:

- Many NP-hard optimization and decision problems have yet to be considered.

- Approximation problems ask for a solution that is within distance δ or ratio ρ from optimal.

- Sharp-P (#P) problems ask for a count of the number of optimal solutions to a given instance.

- Partition problems involve counting the number of solutions for each cost in the solution space.

- Boltzmann (also known as Gibbs) sampling requires to sample from the space of solutions according to a probability distribution that depends on their costs.

- Application areas such as machine learning require solving a large number of hard optimization problems through repeated queries. In other applications, large problems may be decomposed or approximated to obtain hints about optimal solution structures.

Again, *much more work is needed* to understand the full potential of this new approach to problem-solving.

The remainder of this section takes a quick look at experience on two problems that are well-matched to the hardware topology. Subsequent sections consider work on problems that are more challenging to formulate for the D-Wave architecture.

Weighted maximum cut.   For purpose of historical record, we note that Van der Ploeg et al. [106] report on the use of systems of two, three, and four flux qubits in an experimental AQC device: this is the first published report of a successful computation performed by a D-Wave prototype. Prior to this, Steffen et al. [102] had described a successful computation using a three-qubit AQC device built using NMR technology; see Section 4.4 for a brief discussion of that system.

Both devices solved the Weighted Maximum Cut (WMC) problem: Given a graph $G = (V, E)$ with positive edge weights, find a *cut*—a partition of the vertices into nonempty subsets A and B—that maximizes the sum of edge weights across the partition. In the case $n = 3$ (the problem size implemented on these machines), the graph is a triangle and the cut puts one vertex in A and the other two in B: thus the problem is simply to find the pair of edges having maximum sum.

This problem is easy to translate to Ising Model by setting each J_{ij} to equal the weight on edge (i, j), and setting each $h_i = 0$. The positive J_{ij}'s make their incident qubits seek "opposite" spin assignments, and the ground state identifies the pair of edges with largest (negative) sum.

Ising Model. Of course D-Wave platforms can be used to solve the Ising Model. This problem is used in the study of properties of ferromagnetism in statistical mechanics. It is usually defined on a 1, 2, or 3-dimensional lattice, with particles placed on nodes and interaction forces assigned to edges (see Section 3.6 for definition). An initial state is imposed and then the particle system is allowed to evolve to its ground state configuration. Interesting research questions about such systems include: Is there a phase transition from disordered to ordered states? and, What is the configuration of spins in ground state?

Bain et al. [5] describe a test to solve IM instances using an early D-Wave chip built on a C_4 graph with 52 qubits. They measure performance for 300 problem instances with random h_i and J_{ij} taking r values evenly spaced in the range $[-1, +1]$, with $r = 3, 31$ and 511. For example $r = 3$ corresponds to random samples from three values $[-1, 0, +1]$). They observe that performance—in terms of the number of samples needed to find ground state solutions—depends significantly on choice of r. For example, at $r = 1$, mean computation time was near 0.18 seconds, while at $r = 512$, the mean time was close to 0.52 seconds.

5.1.1 TRAINING CLASSIFIERS

In machine learning, a classifier C is an algorithmic device that accepts as input a vector x containing *features*, and assigns to x a *classification* y taken from a small discrete set. A classifier is constructed by an automatic procedure that involves presenting C with a *training set T* containing input and classification pairs (x, y). The goal is for the classifier to "learn" to classify the training set by iteratively self-modifying a collection of weights used in its decision function $c(x) \rightarrow y$. After the training period the classifier is presented with an *evaluation set R* and tested on how well it has learned from T to correctly classify elements of R.

Good classifiers balance accuracy against simplicity and generality: that is, one tolerates a certain level of classification error to avoid overfitting to the training set, which would reduce the usefulness of the classifier when applied to feature sets arising in practice. Classifiers are of interest in many application areas including image recognition, speech recognition, medical diagnosis, natural language understanding, and intrusion detection in networks.

Bain et al. [5] describe use of a D-Wave One chip based on a C_4 chimera graph with 52 working qubits, to construct two types of classifiers. The first is deterministic, computing the single function $c(x) \rightarrow y$. The second is probabilistic, outputing a distribution of classifications.

The optimization problem to be solved has both *hard* and *soft constraints*. Taking their example, suppose the inputs x_i are sentences and the goal is for the classifier to learn to label the parts of speech (noun, verb, etc.) in the sentences. This involves assigning binary labels y_{jp} where $y_{jp} = 1$ means that word j is part of speech p. An example of a hard constraint is: every word must be labeled. An example of a soft constraint is: if word j is a noun, then word $j - 1$ is likely to be an adjective. These rules are encoded in the objective function with high penalties for violations of hard constraints (which correspond to infeasible outcomes) and lower penalties for violations of soft constraints.

Given a set of k training samples $(x, y) \in T$ (for example each x a sentence and each y a bit string of y_{jp} values) the goal is to find a set of weights w to maximize an objective function with two parts: one term $R(w)$ incorporates constraints and minimizes errors on the training set, and one term $\Omega(w)$ favors simple weighting schemes to avoid overfitting to T. In the training phase the learning algorithm iterates to modify the weight set w incrementally. At each iteration a query is sent to the quantum annealing hardware to find an optimal solution to a QUBO instance (see Section 3.2.1) and thereby improve w.

The authors point out that theirs is an unusual objective function because it has quadratic terms, which allows it to account for pairwise correlations between elements of the test set. Their experiments test whether the introduction of a quadratic objective function yields better classifiers. They compare their approach to an alternative method involving support vector machines (SVM) that does not incorporate the quadratic interaction term.

Their evaluation of the classifier looks at two input classes, one with labels based on solutions to random satisfiability problems, and one from a standard test set in image recognition. Maximum instance sizes are $|x| = 20, |y| = 34$ in the random problems and $|x| = 294, |y| = 6$ for the image data. The quality of a learning algorithm is evaluated in terms of *relative Hamming error* which measures the distance between true assignments and those returned the classifier on the evaluation set R. On these tests their learning algorithm produced classifiers with 28 and 10 percent better relative Hamming error than those from the SVM approach.

More discussion of this use of D-Wave quantum annealers to train classifiers may be found in Nevin et al. [76], Nevin et al. [77], and Nevin et al. [78].

5.1.2   FINDING RAMSEY NUMBERS

Suppose you have guest list for a party of N people. For given m and n, you want to determine whether there is a subgroup of size m who are all mutually acquainted, forming a clique, *or* a subgroup of size n who are all strangers, forming an independent set. (If so, your party will be a flop because large cliques and large groups of strangers do not mingle well.)

A *Ramsey number* $R(n, m)$ is a threshold value for N such that if $N \geq R(m, n)$ it must hold that *all* parties of size N will either contain m mutual acquaintances or n mutual strangers. The computation of $R(n, m)$ is a notoriously difficult problem; for example with $n, m > 3$ only nine Ramsey numbers are known. Besides party planning, the problem has important applications in number theory, combinatorics, and complexity theory (see Roberts [90]).

The complexity of this problem is unknown: Schaefer [95] shows that a more general version of the decision problem (for general subgraphs) is co-NP^{NP}-complete, that is, complete for the second level of the polynomial hierarchy.

Gaitan and Clark [41] show that the problem of computing Ramsey numbers belongs in the quantum complexity class QMA (see Section 2.4). They describe a quantum annealing algorithm to compute Ramsey numbers, as follows. The first step is to cast the computation of $R(n, m)$ as an optimization problem. The party list can be represented by a graph G on N vertices and M

edges, where edge (v, w) exists if v and w are acquainted. Given integers n, m we want to know if G contains a clique of size at least m or an independent set of size at least n.

For a given vertex subset $S_\alpha \subseteq V$ of size m, we can compute the product

$$C_\alpha = \prod_{(i > j \in S_\alpha)} a_{ij}. \tag{5.1}$$

which equals 1 exactly when S_α forms an m-clique, and otherwise equals 0. The total number of ways to form m-cliques is found by summing over all such subsets of G:

$$C_m(G) = \sum_\alpha C_\alpha. \tag{5.2}$$

The number $I_n(G)$ of n-independent sets of G can be computed similarly using $I_\alpha = \Pi_{i > j \in S_\alpha}(1 - a_{i\,j})$. This creates a cost function

$$h_{m,n}(G) = C_m(G) + I_n(G) \tag{5.3}$$

equal to the sum of m-cliques and n-independent sets in G. That is, this function is minimized at 0 when G has no m-cliques and no n-independent sets.

The optimization problem is, given (N, m, n), to find an N-vertex graph G^* that minimizes $h_{n,m}(G)$. Note that when N is below the threshold $R(n, m)$, it holds that $\forall G$, $h_{m,n}(G) = 0$. With an algorithm for this minimization problem in hand, $R(n, m)$ may be computed by starting with $N = 1$ and incrementing until $h_{m,n}(G^*) > 0$, at which point the threshold $R(m, n)$ is found.

G can be encoded as an adjacency matrix such that $a_{ij} = 1$ when the edge is present and otherwise $a_{ij} = 0$. The lower triangle of the adjacency matrix forms a bit string of length $L = n(n-1)/2$:

$$g(G) = a_{2,1} \ldots a_{N,1}, a_{3,2} \ldots a_{N,N-1}. \tag{5.4}$$

Now the problem is to find a bit string $g(G)$ to minimize $h_{m,n}(G)$. Associate the bits in $g(G)$ with the computational basis states $\sigma_0^z \otimes \ldots \otimes \sigma_{L-1}^z$ so that string $g(G)$ is identified with superposition state $|g(G)\rangle$, having eigenvalues $h(G)$. It is straightforward to construct the problem Hamiltonian $\mathcal{H}_R$ such that

$$\mathcal{H}_R |g(G)\rangle = h(G) |g(G)\rangle. \tag{5.5}$$

Bain et al. [6] describe the use of this algorithm on a D-Wave system to compute $R(m, 2)$ for $4 \leq m \leq 8$, and $R(3, 3)$. The problem solved in hardware is to find G to minimize the above objective function.

The transformation of this problem to Chimera-structured Ising Model requires the introduction of ancillary qubits, and the minor-embedding requires additional qubits. The largest problem that can be computed on their 102-qubit C_4 chip is $R(8, 2) = 8$, which uses 28 problem qubits, 26 ancillary qubits, and 30 qubits for minor-embedding, totaling 84 qubits. (This problem is small enough to be solved easily by classical computers.)

The search for $R(m, n)$ involves repeated queries to the hardware, incrementing graph size N (vertices) until the returned cost $c = h_{m,n}(G)$ is positive. Since the quantum annealing algorithm is probabilistic, there is a nonzero probability that it will not find the true optimal cost in any given query. This problem is remedied by repeating the anneal some number k times to obtain a sample of solutions.

The authors report that their QA implementation found $R(3, 3)$ and $R(m, 2)$ for $4 \leq m \leq 8$. Computation time included 270 milliseconds for initialization, plus 1ms anneal and 1.5ms readout time per sample. Therefore total time for k samples corresponds to $t = (270 + 2.5k)$ms.

At the largest problem size $N = 8$ the authors observed success probabilities near 64.5 percent. In this case the expected number of samples to find an optimal solution is less than 2, or at most 275ms. The number needed to observe at least one optimal solution with probability at least .99 is at most 5, which works out to at most 282.5ms.

Note that by taking samples it is also possible to attack the (#P) Ramsey enumeration problem, to count the number of distinct graphs of size N for which $h(a) > 0$. Bain et al. [6] report that they found exact correspondence in all but one case between predicted and observed number of graphs found by the hardware.

5.1.3 PROTEIN FOLDING

A lattice protein folding problem involves a sequence of amino acids (called *beads*) connected by peptide bonds (called *strings*) to form a chain. The problem is to embed the chain in a two- or three-dimensional lattice, with beads on vertices and strings on incident edges. This problem has several constraints: for example a chain cannot double back on itself, and two beads cannot occupy the same lattice point. The quality of an embedding is determined by an energy function calculated on the interaction energies between lattice neighbors (including penalty functions for violated constraints). The goal is to find an embedding that minimizes the energy function: this problem is NP-hard (Berger and Leighton [10]).

An embedding can be encoded by a bit string that describes the sequence of turns taken by the chain. For example in a two-dimensional lattice the four possible directions can be encoded as $00 = \downarrow$, $11 = \uparrow$, $01 = \rightarrow$, $10 = \leftarrow$. Thus the embedding 010010 looks like this:

Perdomo-Ortiz et al. [82] report on a project to use a D-Wave One system containing 115 qubits to optimally fold a four-amino-acid and a six-amino-acid sequence. (These instances are small enough to be solved easily by classical algorithms.)

The transformation of this problem to Chimera-structured Ising Model creates an instance that is larger than the number of available qubits. The authors describe two ways to finess this difficulty using a divide-and-conquer approach to partition into smaller subproblems that can be solved in hardware and then stitched together.

They note that, although the problems were solved optimally by hardware, these instances exhibited quite low success probabilities per individual anneal: in the test on the largest subproblem using 81 qubits, only 13 of 10,000 samples were optimal. This appears to be due to the fairly high degree of precision required for specifying edge weights.

5.1.4 GENERAL OPTIMIZATION

The D-Wave hardware takes Chimera-structured Ising Model as its native problem; can it be used to solve other problems? McGeoch and Wang [70] describe experiments to look at performance of a D-Wave Two platform (439 qubits) called V5[1] using instance classes of increasing "distance" from the native problem:

1. Random Ising Model (IM). These instances have random weights $J_{ij} = \pm 1$ assigned to edges in the hardware graph.

2. Random Weighted Maximum 2-SAT instances (WM2SAT) from a recent competition (Argelich et al. [4]). These are essentially general IM problems without Chimera structure and weights specified with 4 bits of precision.

3. Quadratic Assignment Problem (QAP). These instances, downloaded from the QAPLIB repository [QAPLIB], have general structure and unrestricted weights.

The tests considered performance of the D-Wave chip in case (1), and of a hybrid hardware/software solver called Blackbox (a precursor to QSage described in Section 4.1) in cases (2) and (3).

Three classical implementations that represent a variety of solution approaches were also evaluated: AK, a *specialized* solver for WM2SAT that performed well in the above-mentioned competition; TB, a *general* solver, an open source implementation of Tabu search; and CP, the IBM ILOG CPLEX Optimizer, the leading commercial general optimization solver. (A specialized solver targets a specific combinatorial problem, while a general solver reads both an input instance and an objective function. Typically a specialized solver outperforms a general solver on its speciality problem, but adapting a specialized solver to new problems may expose terrible worst-case performance: generality pays the price of efficiency, but efficiency pays the price of robustness. General solvers are normally evaluated according to robustness (best worst case) rather than fastest performance on particular problem classes. They are of interest because of their "plug-

---

[1]This work has an unusual back-story. The authors were retained by D-Wave as consultants to develop experimental procedures for five benchmark tests specified by a potential client. The results described in the paper were from some pre-benchmark pilot studies.

and-play" capability, and because they can be conveniently adapted to messy real-world problem features.)

The solvers were evaluated according to the quality of solutions returned within fixed time limits. Not surprisingly, the D-Wave hardware performed very well on native problems in set (1), always finding best solutions within a 0.5 second timeout, on problem sizes greater than $n = 150$. In test (2), Blackbox tied AK and TB by finding all optimal solutions within the time limits (roughly 30 minutes, determined by a bound on total objective function evaluations). In test (3), Blackbox found best solutions among the four solvers on 27 of 33 instances, although only two of those solutions matched minimum costs published on the repository website.

5.2 IS IT QUANTUM?

A second thread of research has been aimed at the question of whether D-Wave chips exhibit quantum properties, as opposed to carrying out a classical annealing process involving magnetic forces. This question is generally accepted to mean that the qubits should demonstrate entanglement and superposition during the computation. One significant obstacle to such a demonstration is the lack of suitable test apparatus onboard the chips.

Also, given the unusual design approach it is not clear what sort of demonstration might be considered conclusive. For example, the Hamiltonians used in these platforms create a transition from "more quantum" superposition of qubit states at the beginning to "less quantum" classical states at the end of the annealing process. It is also known that entanglement occurs during the middle part of the transition, but not at the beginning and end. (This limited entanglement is sufficient to ensure correct computation in the underlying model.) "Quantum" and "classical" do not form a dichotomy in this framework: perhaps the right question is not *Is it quantum?* but rather *How quantum is it?*

The next few sections review work to address questions of quantumness in D-Wave chips.

5.2.1 QUANTUM ANNEALING VS. THERMAL ANNEALING

Johnson et al. [55] describe experiments to distinguish whether D-wave chips carry out a true quantum annealing process as opposed to a classical thermal annealing process. These tests were run on a single Chimera cell of eight qubits. The Hamiltonian $\mathcal{H}(t)$ carries the cell from the initial Hamiltonian $\mathcal{H}_I$, which imposes an equal superposition of all states, to a final Hamiltonian $\mathcal{H}_P$ that is minimized at the "all equal" states $\uparrow\uparrow\uparrow\uparrow\uparrow\uparrow\uparrow\uparrow$ and $\downarrow\downarrow\downarrow\downarrow\downarrow\downarrow\downarrow\downarrow$. Thus each individual qubit is equally likely to be read as $\uparrow$ or as $\downarrow$ at the end.

From a single qubit's perspective, the problem landscape starts out flat and then gradually transforms into a two-welled landscape corresponding to $|\uparrow\rangle$ and $|\downarrow\rangle$, as shown in Figure 5.1. An *energy barrier* is gradually raised to separate these two wells. When the energy barrier is low a qubit can move freely between the two states whether it is classical or quantum. However, a classical qubit will experience a *freezout time* according to the height of the barrier, which in these experiments is linearly dependent on transition time $t : 0 \rightarrow t_f$. After the freezout time the

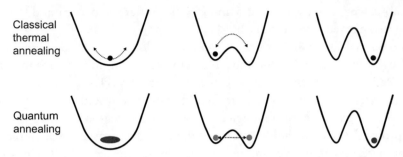

Classical thermal annealing

Quantum annealing

Figure 5.1: The two-welled landscape seen by an individual qubit. The energy barrier between the wells is gradually increased over anneal time $t : 0 \rightarrow T$.

classical qubit does not change state, but a quantum qubit can "tunnel" through the barrier after freezout time, and experience a later freezout that is independent of (or only weakly dependent on) t.

To set up this experiment the authors build k-qubit chains (for $k = 1 \dots 8$) in the cell, with varying weights $J_{ij} = -J$ to create ferromagnetic couplings between chain neighbors (that is, preferring that neighbors have identical spins). The qubits at the two ends of the chain receive strong opposing weights $h_1, h_k = \pm 2J$. This introduces *frustration*, which causes the chain to break somewhere in the middle. If the interior weights $h_2 \dots h_{k-1}$ all equal 0, then the break—called the *domain wall*—will occur between any pair of qubits with equal probability (1/7 at $k = 8$).

The experiment involves changing the interior weights from 0 to a small positive value $h_i = 0.1J$, during the anneal process. This forces the domain wall to be located between qubits 7 and 8. If the qubits respond to this change, the state $|\uparrow\uparrow\uparrow\uparrow\uparrow\uparrow\uparrow\downarrow\rangle$ occurs with much higher probability than 1/7; but if the change occurs after freezout time, classical qubits will have already frozen and will return this state with probability 1/7. Thus changing h_i values at different times during the anneal allows us to detect classical freezout time if it exists.

The authors report on experiments measuring freezout times with varying chain lengths, time scales, and working temperatures. They compare the results to three different numerical simulations involving two models of quantum behavior and one model of classical behavior. At low temperatures $T \leq 50$mK, measured freezout times correlate very well with the more detailed of the two quantum models, and are not consistent with the classical model. At higher temperatures the quantum and classical models converge and both agree with experimental results. This dependence on temperature also supports a quantum explanation of the observations.

5.2.2 DEMONSTRATION OF ENTANGLEMENT

Lanting et al. [64] report on experiments to demonstrate quantum entanglement among two- and eight-qubit sections of a D-Wave chip running at 12.5mK. One set of experiments focuses on pairs of qubits weighted by $J = -2.5$ and $h = 0$, so that the two ground states are $|\uparrow\uparrow\rangle$ and $|\downarrow\downarrow\rangle$, and the two excited states are $|\uparrow\downarrow\rangle$ $|\downarrow\uparrow\rangle$.

It is known that the Hamiltonian $\mathcal{H}(t)$ used in D-Wave systems does not induce entanglement at the beginning and end of the anneal. Entanglement midway means that qubit states are correlated while in superposition, that is, before they arrive at classical states defined by $\mathcal{H}_F$. If there is no entanglement, then an observation mid-transition would yield all four classical states with equal probability; if there is entanglement then the two ground states would be observed with higher probability. It also holds that entanglement is destroyed if the ambient temperature is higher than a certain threshold proportional to the spectral gap, so it is possible to observe changes in entanglement by manipulating ambient temperature T and weights h to change that relationship.

The authors describe experiments that involve measuring the energy distribution of individual qubits. This is based on a novel qubit tunnelling spectroscopy (QTS) technique that uses probe qubits to observe and measure the eigenspectrum and percentage of time spent at each energy level during the anneal process. They also use a technique called *susceptibility-based witnesses* to detect entanglement at ground state. They manipulate temperature T as well as parameters t_f, s, and h while using their experimental apparatus to detect entanglement in mid-transition. They describe four separate sets of evidence that all support an observation of entanglement in two and eight-bit qubit cells, and do not match classical models.

5.2.3   SIGNATURES OF QUANTUM ANNEALING

The previously described experiments focus on small C_1 and C_2 structures and cannot be easily extended to larger hardware graphs. Rather than performing experiments on physical hardware, Boixo et al. [14] and Boixo et al. [15] take an indirect approach to consider the question of quantumness at larger scales, by looking at solution *signatures* of a D-Wave One chip compared to those of classical software solvers. The signature of a solver with respect to an input class I is a property defined on the empirical distribution of instances the solver found easy versus hard to solve.

This approach sets up a sort of Turing Test for quantumness: as the argument goes, if the signature of the D-Wave hardware resembles that of a quantum model then we have evidence that a quantum computation occurred; but if the signature resembles a classical model of computation we have evidence of classical behavior. Ideally the signature would resemble one model and *not* the other. Formal criteria for "resemblance" in this context are not available.

Boixo et al. [14] look at small constructed instances that allow detailed analysis of solution signatures. They observe that hardware signatures for a D-Wave One platform are consistent with models of quantum annealing and inconsistent with a models of classical annealing. Boixo et al. [15] describe experiments using a D-Wave One system (DW) to solve random Chimera-structured Ising Model instances. They compare DW signatures to those of simulated quantum annealing (SQA), simulated annealing (SA), and spin dynamics (SD). The first models a quantum annealing process and the latter two model classical processes. The authors observe strong bimodality in QA success probabilities at the largest problem size $n = 108$: that is, instances were either very easy or very difficult to solve in most cases. This bimodality is also strong in the SQA results, and observable but less strong for SD. There is no bimodality in the SA results, which suggests that the hardware computation does not resemble classical thermal annealing. The authors also looked at correlations in success probabilities on individual instances and found stronger correlations between DW and SQA than between the two classical models.

5.3   HOW FAST IS IT?

Table 4.1 in Section 4.1 shows how total computation time can be broken into three components: chip programming time T_p, anneal time T_a, and readout time T_r. The total time to sample k solutions for a given instance is $T_k = T_p + k(T_a + T_r)$.

In current chip models, classical overhead costs dominate computation time. Quantum annealing takes around 20 microseconds, but the total time to solve any given instance is bounded *below* by $T_p + T_a + T_r$ when $k = 1$, which corresponds to hundreds of milliseconds. Classical solvers (with time performance increasing from 0 milliseconds at $n = 0$) can always outperform the hardware on "small enough" instances. Comparisons of time performance between quantum and classical solvers become interesting when n is "large enough" that a crossover point may be observed (if it exists).

The quality of solution returned by a chip in any given anneal run depends on factors relating to the *input* (such as problem size), the quantum annealing *platform* (temperature, noise, and ICE), and the quantum annealing *algorithm* (Hamiltonians and path function), as well as on any error-correction techniques that may be applied. Very little is known about how these factors affect the outcome of a single anneal. Thus the underlying performance model describes a probabilistic tradeoff between time (determined by anneal time and samples k) and solution quality. Successive chip models have yielded reductions in overhead costs T_p and T_r, as well as better error suppression and greater range and precision in weights; this trend is expected to continue. This means that empirical performance results are very difficult to model and to generalize, and very soon obsolete.

Within this context of uncertain conclusions and limited scope, a few reports comparing D-Wave quantum annealing chips to classical solution methods have appeared. The experiments by McGeoch and Wang [70] described in Section 5.1.4 look at the quality of solutions returned by the V5 chip to those returned by three classical solvers, all running within 0.5 second time-outs ($k = 1,000$) on random native instances with $J_{ij} \pm 1$ and $h_i = 0$. At larger problem sizes $150 \leq n \leq 439$, the hardware solutions were best in nearly 100 percent of instances while software success rates were near 0 percent. The best of the classical solvers in that short time limit, AK, could almost match hardware solution quality on problem sizes below $n = 150$. CPLEX, the best of the three classical solvers at longer time limits, was able to match hardware solutions running up to 30 minutes at largest problem sizes.

After this work was published, additional experimental work using the same or similar input classes was reported. Puget [84] found an alternative problem formulation to speed up CPLEX by about 40x on a single core and by larger factors using a multithreaded system. Also, the Google team at QuAIL reported that their tests using a bigger and faster D-Wave chip yielded a 35,500x since speedup over (the earlier formulation of) CPLEX [85].

Others have developed classical solvers specialized to Chimera-structured Ising Model instances. Instead of measuring solution quality with fixed time bounds, these authors look at estimates of *time to solution*, that is the time needed to find an optimal solution with high probability. The experiments by Boixo et al. [15] described in Section 5.2.3 compare performance of a D-Wave One (DW) platform (108 qubits) to their implementations of simulated annealing (SA) and simulated quantum annealing (SQA). They observe similar times between DW and SW at largest problem size $n = 108$. Using a different cost metric, Selby [97] reports times in the same general range as those for the largest problem sizes on V5 and V6 chips described in McGeoch and Wang [70].

Defining and detecting quantum speedup.   The work in Boixo et al. [15] is extended in Rønnow et al. [91], which compares time to solution (TTS) for a D-Wave Two platform (503 qubits) to those of SA and SQA. To look at scaling performance they compare a hardware anneal times to an idealized measure of "work" on a classical platform (discussed below). On random instances with $J_{ij} = \pm 1$ and with $J_{ij} \in [-7 \ldots +7]$ they observe that SA scales better than DW in the sense that the time curve $t(SA)/t(DW)$ is concave in the median and higher quantiles.

The authors are careful to point out that their study should not be construed as addressing larger questions about scaling performance of the quantum annealing hardware in other scenarios. Indeed, they note that the minimal anneal time $T_a = 20$ microseconds was used throughout; lower time settings, likely optimal in the range of problem sizes tested, were not available on the device. Since it was not possible to find the optimal scaling curve for the quantum annealing hardware as was done for the software, extrapolation of the ratio curve to larger problem sizes is not possible.

Also, Katzgraber et al. [59] point out that the class of random instances used in these tests is not suitable for distinguishing DW performance from that of SA or SQA. These inputs lack sufficient structure to be challenging to annealing-based solvers: that is, at nonzero temperatures (in annealing schedules), there are no hills in the solution landscape. This means that SA and SQA are simply working as greedy samplers, and the hardware has no opportunity to exploit its quantum nature by tunnelling through barriers. Katzgraber et al. [59] propose some more interesting test cases that await future analysis.

It is hard to tell whether these results have any implications for performance in practice. For the classical solvers the authors define *reference times* that correspond to the instruction cost of flipping the sign of a single node; for example an Intel Xeon core can perform 5 bit-flips per nanosecond. To separate "quantum" from "parallel" sources of speedup, they divide reference times by n, arguing that a hypothetical purpose-built classical annealing platform could perform

n bit flips in the same time as one bit flip on a single-core machine. Since these reference times ignore low-order instruction costs, memory access, and parallel overhead, they are unlikely to be reflective of computation time in practice. In a separate comparison of wall-clock times the authors observe that SA time-to-solution is generally lower than that for DW, but point out that DW times are dominated by the constant initialization cost T_p.

5.4   EPILOGUE

The theme of this chapter is clear: much more experimental (and theoretical) work is needed before the capabilities of this new approach to computation can be fully understood.

Perhaps that work can someday be generalized to address more fundamental questions: is quantum computation faster than classical computation? But this author is skeptical that such experiments can be devised. Consider, for example, the viability of finding experimental demonstration (or refutation) of a conjecture that $P \neq NP$. The formal statement of the question contains alternating quantifiers ($\exists \forall \exists \ldots$). Experimental methods can sometimes be used to demonstrate the *existence* of a property, but they are not generally useful for demonstrations of universality or non-existence.

The empirical performance of a given optimization algorithm—quantum or classical— depends on interactions between the program, the platform, and the input instance (or input class) on which it is tested. Extending observations about performance of classical algorithms much beyond experimental boundaries is a hazardous undertaking: see McGeoch et al. [68] or McGeoch [69] for tales of extrapolation gone awry. The difficulty is exacerbated when quantum annealers are added to the mix, due to lack of underlying models of what drives performance.

Of course the fact that experiments may be limited in scope is no reason to abandon the knowledge-gathering effort. Empirical work has so far shown that D-Wave's approach to quantum computation is viable, which opens up exciting new directions for bringing quantum computation to bear in real-world applications. A great deal of information and insight can be gained from formulating, implementing, and analyzing algorithms running on D-Wave platforms. This information not only suggests new and better solution methods, but also provides valuable insights about improved architectures and technologies to support the quantum adiabatic computation model.

Or, as Alan Turing put it:

> We can only see a short distance ahead, but we can see plenty there that needs to be done.

Bibliography

[1] D. Aharonov, W. van Dam, J. Kempe, Z. Landau, S. Loyd, and O. Regev "Adiabatic quantum computation is equivalent to standard quantum computation." *SIAM Journal of Computing*, vol. 37, issue 1, pp. 166–194, 2007. The conference paper appeared in *Proceedings of 45th FOCS, pp. 42–51*, 2004. DOI: 10.1137/S0097539705447323. 3, 18, 26, 27

[2] B. Altshuler, H. Krovi and J. Roland, "Anderson localization makes adiabatic quantum optimization fail," *Proceedings of the National Academy of Science, USA* 107(28), pp. 12446–12450. DOI: 10.1073/pnas.1002116107. 22

[3] B. Apolloni, C. Carvalho, and D. de Falco, "Quantum stochastic optimization," *Stochastic Processes and their Applications* 33 pp. 233–244, 1989. DOI: 10.1016/0304-4149(89)90040-9. 29

[4] Joesph Argelich, Chu Min Li, Felip Manyá and Jordi Planes (organizers), Max-SAT 2012: Seventh Max-SAT Evaluation, `maxsat.udl.cat`. The testbed cited here may be found in the Weighted Max-SAT/Random category. 65

[5] Zhengbing Bain, Fabian Chudak, William G. Macready, and Geordie Rose, "The Ising model: teaching an old problem new tricks," D-Wave Systems Technical Report August 30, 2010. Available from `/www.dwavesys.com/resources/publications`. 39, 40, 49, 61

[6] Zhengbing Bain, Fabian Chuduk, William G. Macready, Lane Clark, and Frank Gaitan, "Experimental determination of Ramsey numbers with quantum annealing," Physical Review Letters Vol 111, 130505 (2013). DOI: 10.1103/PhysRevLett.111.130505. 63, 64

[7] Nikhil Bansal, Sergey Bravyi, and Barbara M. Terhal, "Classical approximation schemes for the ground-state energy of quantum and classical Ising spin Hamiltonians on planar graphs," *Journal of Physics A*, Vol 15, No 10, 1982. 37

[8] Demian A. Battaglia, Giuseppe E. Santoro, and Erio Tosatti, "Optimization by Quantum Annealing: Lessons from hard 3-SAT cases," *Phys. Rev. E* 71, 066707, 2005. DOI: 10.1103/PhysRevE.71.066707. 35, 36, 41

[9] Demian Battaglia, Lorenzo Stella, Osvaldo Zagordi, Giuseppe E. Santoro, and Erio Tosatti, *Deterministic and stochastic quantum annealing approaches*, Springer Lecture Notes in Physics 679, pp. 171–206, 2005. DOI: 10.1007/11526216_7. 35, 36, 38

[10] B. Berger and T. Leighton, "Protein folding in the hydrophobic-hydrophilic (HP) model is NP-Complete," *J. Comput. Biol.* vol 5, pp. 27–40, 1998. DOI: 10.1089/cmb.1998.5.27. 64

[11] A. J. Berkeley, M. W. Johnson, P. Bunyk, R. Harris, J. Johannson, T. Lanting, E. Ladizinsky, E. Tolkacheva, M. H. S. Amin and G. Rose, "A scalable readout system for a superconducting adiabatic quantum optimization system," *Superconductivity Sci. Technol.* 23 (2010) 105014, 2010. DOI: 10.1088/0953-2048/23/10/105014. 53

[12] Dimitris Bertsimas and John Tsitsiklis, "Simulated Annealing," *Statistical Science*, Vol 8, No 1, pp. 10–15, 1993. DOI: 10.1214/ss/1177011077. 33

[13] Jacob D. Biamonte and Peter J. Love, "Realizable Hamiltonians for universal adiabatic quantum computers," *Phys. Rev A* 78, 012352 (2008). DOI: 10.1103/PhysRevA.78.012352. 27

[14] Sergio Boixo, Tameem Albash, Frederico M. Spedalieri, Nicholas Chancellor, and Daniel A. Lidar, "Experimental signature of programmable quantum annealing," *Nature Communications*, 4, Ar ticle 2067, 28 June 2013. DOI: 10.1038/ncomms3067. 54, 55, 68

[15] Sergio Boixo, Troels F. Rønnow, Sergei V. Isakov, Zhijui Wang, David Wecker, Daniel A. Lidar, John M. Martinis and Matthias Troyer, "Evidence for quantum annealing with more than one hundred qubits," *Nature Physics* 10, pp. 218–224, 28 February 2014. DOI: 10.1038/nphys2900. 55, 68, 70

[16] M. Born and V. Fock "Beweis des Adiabatensatzes," *Zeitschrift für Physik A*, 51 (3–4) pp. 165–180. DOI: 10.1007/BF01343193. 5, 18

[17] Endre Boros, Peter L. Hammer, and Gabriel Tavares, "Local search heuristics for Quadratic Unconstrained Binary Optimization (QUBO)," *J. Heuristics*, 2007, 13, pp. 99–132. DOI: 10.1007/s10732-007-9009-3. 45

[18] Sergey Bravyi, David P. DiVincenzo, Roberto I. Oliveeira, and Barbara M. Terhal, "The complexity of stoquastic local Hamiltonian problems," *Quant. Inf. Comp.* Vol 8 No 5, pp. 361–385. 2008 26

[19] P.I. Bunyk et al., "Architectural considerations in the design of a supercomputing quantum annealing processor," 2014. In submission. 43, 52, 53

[20] Peter Cheeseman, bob Kanefsky, and William M. Taylor, "Where the really hard problems are," *Proceedings of the 12th International Joint Conference on AI (IJCAI-91)*, Morgan Kaufman, Vol 1, pp. 331–340, 1991. 21

[21] A. M. Childs, E. Farhi, and J. Preskill, "Robustness of adiabatic quantum computation," *Phys. Rev. A* 65, 012322. 2001. DOI: 10.1103/PhysRevA.65.012322. 53, 56

[22] Andrew Childs, Lecture 19: The quantum adiabatic theorem, *Quantum Algorithms* (CO 871 Winter 2008) 18

[23] Adrian Cho, "Controversial computer is at least a little quantum mechanical," *Science* 13 May 2011.

[24] Vicky Choi, "Minor-embedding in adiabatic quantum computation: I. The parameter setting problem," *Quantum Information Processing*, Vol 7 Issue 5, October 2008, pp. 193–209, 2008. DOI: 10.1007/s11128-008-0082-9. 47

[25] Vicky Choi, "Minor-embedding in adiabatic quantum computation: II. Minor-universal graph design," *Quantum Information Processing*, Vol 10 Issue 3, June 2011, pp. 343–353. DOI: 10.1007/s11128-010-0200-3. 47

[26] Vicky Choi, "Different adiabatic quantum optimization algorithms for the NP-complete exact cover and 3SAT problems," Arxiv May 2011. DOI: 10.1073/pnas.1018310108. 22

[27] Arnab Das and Bikas K. Chakrabarti, eds. *Quantum Annealing and Related Optimization Methods*, Springer Lecture Notes in Physics 679, 2005. DOI: 10.1007/11526216. 31, 35, 81

[28] "Quantum theory, the Church-Turing principle and the universal quantum computer," *Proceedings of the Royal Society A* 400 (1818), pp. 97–117, 1985. DOI: 10.1098/rspa.1985.0070. 3, 13

[29] David Deutsch and Richard Jozsa, "Rapid solutions of problems by quantum computation," *Proceedings of the Royal Society of London A* pp. 439–552, 1992. DOI: 10.1098/rspa.1992.0167. 4

[30] Neil G. Dickson and Mohammed H. Amin, "Algorithmic approach to adiabatic quantum optimization," *Physics Review A* 85.032303 2012. DOI: 10.1103/PhysRevA.85.032303. 23

[31] NG Dickson, MW Johnson, MH Amin, R Harris, F Altomare, AJ Berkely, P Bunyk, J Cai, EM Chapple, P Chavez, F Ciota, T Cirip, P Debuen, M Drew-Brook, C Enderud, S Gildert, F Hamze, JP Hilton, E Hoskinson, K Karimi, E Ladizinsky, N Ladizinsky, T Lanting, T Mahon, R Neufeld, T Oh, I Perminov, C Petroff, A Przbysz, C Rich, P Spear, A Tcaciuc, MC Thom, E Tolkacheva, S Uchaikin, J Wang, AB Wilson, Z Merali and G. Rose, "Thermally assisted quantum annealing of a 16-qubit problem," *Nature Communications*, 4:1903, 2013. DOI: 10.1038/ncomms2920. 54

[32] Jaifeng Du, Nanyan Xu, Xinhua Peng,Pengfei Wang, Sanfeng Wu, and Dawei Lu, "NMR Implementation of a molecular hudrogen quantum simulation with adabatic state preparation," *Physical Review Letters*, 105, 030502 January 2012. DOI: 10.1103/PhysRevLett.104.030502.

[33] Stefan Edelkamp and Stefan Schrödl, *Heuristic Search: Theory and Applications*, Morgan Kaufman/Elsevier, 2012. 32

[34] E. Farhi, J. Goldsone, D. Gosset, S. Gutmann, H. B. Meyer, and P. Shor, "Quantum adiabatic algorithms, small gaps, and different path s," MIT CPT 4076 and CERN-ph-th-2009/175. *Quantum Information and Computation*, Vol 11 Issue 3, March 2011. pp. 181–214. 22

[35] E. Farhi, J. Goldstone, S. Gutmann, and M. Sipser "Quantum computation by adiabatic evolution." arXiv:quant0ph/00001106v1. 2000 2, 3, 9, 16, 21, 22

[36] E. Farhi, J. Goldstone, S. Gutmann, J. Lapan, A. Lundgren, and D. Preda "A quantum adiabatic evolution algorithm applied to random instances of an NP-Complete problem." *Science,* vol. 292, no. 5516, pp. 472–475, 20 April 2001. DOI: 10.1126/science.1057726. 2, 9, 19, 21, 26, 35

[37] Richard Feynman, talk delivered at the *First Conference on the Physics of Computation*, MIT May 1981. See Feynman, "Simulating physics with computers," *Int. J. of Theoretical Physics*, V 21, No 6/7, pp. 467–488. DOI: 10.1007/BF02650179. 3

[38] A. B. Finnila, M. A. Gomez, C. Sebnik, C. Stenson, and J. D. Doll, "Quantum Annealing: A new method for minimizing multidimensional functions," *Chem. Phys. Letters* 219, pp. 343–348, 1994. DOI: 10.1016/0009-2614(94)00117-0. 29

[39] Y. Fu and P. W. Anderston, "Applications of statistical mechanics to NP-complete problems in combinatorial optimisation," *J. Phys. A*, 19(9), pp. 1605–1620, 1986. DOI: 10.1088/0305-4470/19/9/033. 37

[40] Michael Garey, David S. Johnson, and Larry Stockmeyer, "Some simplified NP-complete graph problems," *Theor. Comput. Sci.* 1:237–267, 1976. DOI: 10.1016/0304-3975(76)90059-1. 39

[41] Frank Gaitan and Lane Clark, "Ramsey numbers and adiabatic quantum computing," *Physics Review Letters* 108, 010501 (2012) DOI: 10.1103/PhysRevLett.108.010501. 35, 62

[42] Google Quantum AI Lab: Blog post: Where do we stand on benchmaring the D-Wave 2? January 19, 2014. *plus.google.com/+QuantumAILab/posts* accessed 5/2014.

[43] G. Greenstein and A. G. Zijonc, *The Quantum Challenge* Jones and Bartlett Pub., Sudbury MA, 1997. 9

[44] Lev Grossman, "The quantum quest for a revolutionary computer," *Time Magazine*, Monday Feb. 17, 2014. 7

[45] L. K. Grover, "A fast quantum mechanical algorithm for database search," *Proceedings*, 28th STOC, p. 212, May 1996. DOI: 10.1145/237814.237866. 3

[46] Stuart Geman and Donald Geman, "Stochastic relaxation, Gibbs distributions, and the Bayesian restoration of images," *IEEE Transactions on Pattern Anal. Mach. Intell.*, 6, pp. 721–741, 1984. DOI: 10.1109/TPAMI.1984.4767596. 33

[47] Lev Grossman, "The Quantum Quest for a Revolutionary Computer," *Time* Feb 6, 2014.

[48] R. Harris, M. W. Johnson, T. Lanting, A. J. Berkley, J. Johansson, P. Bunyk, E. Tolkacheva, E. Ladizinsky, T. Oh, F. Cioata, I Perminov, P. Spear, C. Enderud, C. Rich, S. Uchaikin, M. C. Thom, E. M. Chapple, J. Wang, B. Wilson, M. H. S. Amin, N. Dickson, K. Karimi, B. Macready, C. J. S. Truncik, and G. Rose, "Experimental investigation of an eight-qubit unit cell in a superconducting optimization procoessor," *Phys. Rev. B* 82 (2):024511, July 2010. DOI: 10.1103/PhysRevB.82.024511. 43

[49] R. Harris, J. Johansson, A. J. Berkley, M. W. Johnson, T. Lanting, S. Han, P. Bunyk, E. Ladizinsky, T. Oh, I. Perminov, E. Tolkacheva, S. Uchaikin, E. Chapple, C. Enderud, C. Rich, M. Thom, J. Wang, B. Wilson, G. Rose, "Experimental demonstration of a robust and scalable flux qubit," *Phys. Rev B.* 81, 124510 (2010). DOI: 10.1103/PhysRevB.81.134510. 43, 48, 50

[50] Itay Hen and A. P. Young, "Exponential complexity of the quantum adiabatic algorithm for certain satisfiability problems," Arxiv 1109.6872v2, 2011. DOI: 10.1103/PhysRevE.84.061152. 23, 40

[51] Tad Hogg, "Adiabatic quantum computing for random satisfiability problems," *Phys Rev A* 67, 2003. DOI: 10.1103/PhysRevA.67.022314. 22

[52] Jeremy Hsu, "D-Wave's Year of Computing Dangeriously," *IEEE Spectrum*, 26 November 2013. DOI: 10.1109/MSPEC.2013.6676982. 7

[53] Sorin Istrail, "Statistical mechanics, three-dimensionality and NP-completeness: I. Universality of intractability for the partition function of the Ising model across non-planar surfaces," *Proceedings of STOC '00*, pp. 87–96, 2000. DOI: 10.1145/335305.335316. 37

[54] Thomas Jansen, "Introduction to the theory of complexity and approximation algorithms," in Ernst W. Mayr et al., *Lectures on Proof Verification and Approximation Algorithms*, Springer pp. 5–28. DOI: 10.1007/BFb0053011. 25

[55] M. W. Johnson, M. H. S. Amin, S. Gildert, T. Lanting, F. Hamze, N. Dickson, R. Harris, A. J. Berkley, J. Hohansson, P. Bunyk, E. M. Chapple, C. Enderud, J. P. HIlton, K. Karimi, E. Ladizinsky, N. Ladizinsky, T. Oh, I. Perminov, C. Rich, M. C. Thom, E. Tolkacheva, C. J. S. Truncik, S. Uchaikin, J. Wang, B. Wilson, and G. Rose, "Quantum annealing with manufactured spins," *Nature* Vol 473, 12 May 2011. DOI: 10.1038/nature10012. 43, 66

[56] Tadashi Kadowaki and Hidetoshi Nishimori, "Quantum annealing in the transverse Ising model," *Phys. Rev. E* 58, pp. 5355–5363. 1998. DOI: 10.1103/PhysRevE.58.5355. 29, 31, 38

[57] Alan Kay, remarks at a 1971 meeting, Xerox PARC. 7

[58] Kamran Karimi, Neil G. Dickson, Firas Hamze, M. H. S. Amin, Marshal Drew-Brook, Fabian A. Chudak, Paul I. Bunyk, William G. Macready, and Georgie Rose, "Investigating the performance of an adiabatic quantum optimization processor," arXiv:1006.4147v4, 27 January 2011. DOI: 10.1007/s11128-011-0235-0.

[59] Helmut G. Katzgraber, Firas Hamze, and Ruben S. Andrist, "Glassy Chimeras could be blind to quantum speedup: Designing better benchmarks for quantum annealing machines," ArXiv 1401.1546 12 January 2014. DOI: 10.1103/PhysRevX.4.021008. 70

[60] Phillip Kaye, Raymond Laflamme, and Michele Mosca, *An Introduction to Quantum Computing*, Oxford University Press, 2007.

[61] S. Kirkpatrick, C. D. Gelatt, and M. P. Vecchi, "An approach to optimization based on simulated annealing", *Science* 220, pp. 671–680, 1983. DOI: 10.1126/science.220.4598.671. 33

[62] P. J. van Laarhoven and E. H. Aarts, *Simulated Annealing: Theory and Applications*, Springer series on Mathematics and its Applications No 37, 1987. DOI: 10.1007/978-94-015-7744-1. 33

[63] "Quantum Computing," *Nature* 464, pp. 45–53, 2010. 50

[64] T. Lanting, A. J. Przbysz, A. Yu. Smirnov, F. M. Spedalieri, M. H. Amin, A. J. Berkely, R. Harris, F. Altomare, S. Boixo, P. Bunyk, N. Dickson, C. Enderud, J. P. HIlton, E. Hoskinson, M. W. Johnson, E. Ladizinsky, N. Ladizinsky, R. Neufeld, T. Oh, I. Perminov, C. Rich, M. C. Thom, E. Tolkacheva, S. Uchaikin, A. B. Wilson, and G. Rose, "Entanglement in a quantum annealing processor," arXiv:1401.3500v1 15 January 2014. DOI: 10.1103/PhysRevX.4.021041. 67

[65] F. S. Levin, *An Introduction to Quantum Theory*, Cambridge: Cambridge University Press, 2002.

[66] Roman Martoňák, Guiseppe E. Santoro, and Erio Tosatti, "Quantum annealing of the traveling salesman problem," *Physical Review E 70*, 057701, 2004. 35

[67] Y. Matsuda, H. Nishimori, and H. G. Katzgraber, "Ground-state statistics from annealing algorithms: Quantum versus classical approaches," *New Journal of Physics* 11:073021, 2009. DOI: 10.1088/1367-2630/11/7/073021. 35

[68] Catherine McGeoch, Peter Sanders, Rudolf Fleischer, Paul R. Cohen, and Doina Precup, "Using finite experiments to study asymptotic performance," Dagstuhl seminar on experimental algorithmics, Springer LNCS 2547, pp. 93–136, 2002. 71

[69] Catherine C. McGeoch, *A Guide to Experimental Algorithmics*, Cambridge University Press, 2011. DOI: 10.1017/CBO9780511843747. 71

[70] Catherine C. McGeoch and Cong Wang, "Experimental evaluation of an adaiabatic quantum system for combinatorial optimization," *Proceedings of the 2013 Conference on Computing Frontiers*, ACM 2013. DOI: 10.1145/2482767.2482797. 49, 65, 69, 70

[71] N. David Mermin, *Quantum Computer Science: An Introduction*, Cambridge University Press, 2007. DOI: 10.1017/CBO9780511813870. 9

[72] David Mitchell, Bart Selman, and Hector Levesque, "Hard and easy distributions of SAT problems," Proceedings of the Tenth National Conference on Artificial Intelligence (AAAI-92), 1992. 21

[73] Zbigigniew Michalewicz and David B. Fogel, *How to Solve It: Modern Heuristics*, Springer Verlag 2004. DOI: 10.1007/978-3-662-07807-5. 32, 33

[74] Satoshi Morita and Hidetoshi Nishimori, "Mathematical foundation of quantum annealing," *Journal of Mathematical Physics*, 49, 125210, 2008. DOI: 10.1063/1.2995837. 31, 35

[75] Michael A Nielsen and Isaac L. Chuang *Quantum Computation and Quantum Information: 10th Anniversary Edition*, Cambridge University Press, 2011. 3, 9, 14

[76] Hartmut Nevin, Geordie Rose, and William G. Macready, "Image recognition with an adiabatic quantum computer I: Mapping to quadratic unconstrained binary optimization," arXiv:0804.4457, 28 April 2008. 62

[77] Hartmut Nevin, Vasil S. Denchev, Geordie Rose and William G. Macready, "Training a binary classifier with the quantum adiabatic algorithm," arXiv:0811.0416, 4 Nov. 2008. 62

[78] Hartmut Nevin, Vasil S. Denchev, Geordie Rose and William G. Macready, "Training a large scale classifier with the Quantum adiabatic algorithm," arXiv:0912.0779, 4 December 2009. 62

[79] "Hope for new data in air photo of sun," *New York Times* pp. 17. May 15, (1919). 1

[80] Roberto Oliveira and Barbara M. Terhal, "The complexity of quantum spin systems on a two-dimensional square lattice," *Quant. Inf. Comp.* 8(10), pp. 900–924. 37

[81] Masayuki Ohzeki and Hidetoshi Nishimori, "Quantum annealing: An introduciton and new developments," *Journal of Computational Theor. Nanoscience*, 8 (2011) p. 963. DOI: 10.1166/jctn.2011.1776963. 35

[82] Alejandro Perdomo-Ortiz, Neil Dickson, Marshall Drew-Brook, Geordie Rose, and Alán Aspuru-Guzik, "Finding low-energy conformations of lattice protein models by quantum annealing," *Nature Scientific Reports*, Vol 2, Artcile 571, August 2012. DOI: 10.1038/srep00571. 64

[83] Kristen L. Pudenz, Tameen Albash, and Daniel A. Lidar "Error-corrected quantum annealing with hundreds of qubits," *Nature Communications*, 5, Article 3243, 6 February 2014. DOI: 10.1038/ncomms4243. 55

[84] Jean Francois Puget, *IT Best Kept Secret is Optimization* Blog post: Is quantum computing useful for optimization? May 17, 2013. `www.ibm.com/developerworks/community/blo gs/jfp/?lang=en` keyword search on D-Wave. Accessed 5/2014. 70

[85] Google Quantum A. I. Lab Team, Blog post: Where do we stand on benchmarking the D-Wave 2? January 19, 2014. *plus.google.com/+QuantumAILab/posts/DymNo8DzAYi*. Accessed 5/2014. 70

[QAPLIB] R. E. Burkard, E. Çela, S. E. Karisch, and F. Rendl, QAPLIB Quadratic Assignment Problem Library: Problem Instances and Solutions. `www.seas.upenn.edu/quaplib/ins t.html` 65

[87] P. Ray, B. K. Chakrabarti, and A. Chakrabarti, "Sherrington-Kirkpatric model in a transverse field: Absence of replica symmetry breaking due to quantum fluctuations," *Phys. Rev. B* 39 11828 (1989) DOI: 10.1103/PhysRevB.39.11828. 29

[88] E. Rieffel and W. Polak "An introduction to quantum computing for non-physicists." *ACM Computing Surveys* vxx nxx 2001 DOI: 10.1145/367701.367709. 3, 4, 9, 14

[89] Ben W. Reichardt "The quantum adiabatic optimization algorithm and local minima," *ACM STOC*, Chicago 2004. DOI: 10.1145/1007352.1007428. 18

[90] Fred S. Roberts, "Applications of Ramsey Theory," *Discrete Applied Mathematics*, 9, pp. 251–261. 1984. DOI: 10.1016/0166-218X(84)90025-8. 62

[91] Troels Rønnow, Zhihui Wang, Joshua Job, Sergio Boixo, Sergei V. Isakov, David Wecker, John M. Martinis, Daniel A. Lidar, and Matthias Troyer, "Defining and deteting quantum speedup," arXiv: 1401.2910v1, 13 January 2014. DOI: 10.1126/science.1252319. 70

[92] G. Rose and W. G. Macready, "An introduction to quantum annealing," DWAVE/Technical Document 0712 availlable at `dwave.files.wordpress.com/.. ./20070810_d-wave_quantum_annealing.pdf`. DOI: 10.1051/ita/2011013. 29

[93] Rishi Saket, "A PTAS for the classical Ising spin problem on the chimera graph structure," arXiv: 1306.6943v2, July 2013. 37

[94] Giuseppe E. Santoro and Erio Tosatti, "Optimization using quantum mechanics: quantum annealing through adiabatic evolution," *Journal of Physics A: Mathematical and General* V 39, No 36. 2008 DOI: 10.1088/0305-4470/39/36/R01.

[95] Marcus Schaefer, "Graph Ramsey theory and the polynomial hierarchy," *Journal of Computer and System Sciences*, V62, Issue 2, March 2001, pp. 290–322. DOI: 10.1006/jcss.2000.1729. 62

[96] Gernot Schaller and Ralf Schützhold, "The role of symmetries in adiabatic quantum algorithms," *Quantum Information and Computation* 10, 0109–0140 (2010). 22, 41

[97] Alex Selby, Blog post: D-Wave: comment on comparison with classical computers (and subsequent posts), www.archduke.org/stuff/d-wave-comment-on-comparison-with-classical-computers/. Accessed May 2014. 70

[98] Parongama Sen and Pratap K. Das, "Dynamical Frustration in ANNNI Model and Annealing," in Das and Chakrabarti [27], pp. 325–337. DOI: 10.1007/11526216_12. 38

[99] Peter Shor, "Algorithms for quantum computation: Discrete logarithms and factoring," *FOCS 1994* Santa Fe, 1994. DOI: 10.1109/SFCS.1994.365700. 3

[100] Peter Shor, "Polynomial-time algorithms for prime factorization and discrete logarithms on a quantum computer," *SIAM Journal of Computing*, 26, pp. 1484–1509, 1997. DOI: 10.1137/S0097539795293172.

[101] Rolando D. Somma and Sergio Boixo, "Spectral Gap Amplfiication," *SIAM Journal on Computing*, 42 (2013) pp. 593–610. DOI: 10.1137/120871997. 23

[102] Matthias Steffen, Wim van Dam, Tad Hogg, Greg Breyta, and Isaac Chuang, "Experimental implementation of an adiabatic quantum optimization algorithm," *Physical Review Letters*, Vol 90, No 6, Feb. 2003. DOI: 10.1103/PhysRevLett.90.067903. 56, 60

[103] Gabriel Tavares, *New Algorithms for Quadratic Unconstrained Binary Optimization (QUBO) with Applications in Engineering and Social Sciences*, PhD Disseration in Operations Research, Rutgers University, New Brunswick, NJ 2008. 39, 45

[104] D. Thirumalai, Q. Li, and T. R. Kirkpatrick, "Infinite range Ising spin glass in a transverse field," *J. Phys.* A 22, pp. 3339–3349, 1989. DOI: 10.1016/0375-9601(87)90190-3. 29

[105] Wim van Dam, Michele Mosca, and Umesh Vazirani, "How powerful is adiabatic quantum computation?" *Proceedings 42nd FOCS*, pp. 279–287, 2001. DOI: 10.1109/SFCS.2001.959902. 18, 26

[106] S. H. W. van der Ploeg, A. Izmalkov, M. Grajcar, U. Hübner, S. Linzen, S. Uchaikin, Th. Wagner, A. Yu. Smirnov, A. Maassen van den Brink, M. H. S. Amin, A. M. Zagoskin, E. Il'ichev and H.-G. Meyer, "Adiabatic quantum computation with flux qubits, first experimental results," *Applied Superconductivity Conference*, 2006. DOI: 10.1109/TASC.2007.898156. 60

[107] Rodney Van Meter and Clare Horsman, "A blueprint for building a quantum computer," *Communications of the ACM*, Vol 56, No 10, pp. 84–93, 2013. DOI: 10.1145/2494568. 5

[108] Colin P. Williams, *Explorations in Quantum Computing*, Springer, 2011. DOI: 10.1007/978-1-84628-887-6. 9

[109] Nianyan Xu, Jin Zhu, Dawei Lu, Xianyi Zhou, Xinhua Peng, and Jiangfen Du, "Quantum factorization of 143 on a dipolar-coupling nuclear magnetic resonance system," *Physical Review Letters 108*, Issue 13, 130501 30 March (2012). DOI: 10.1103/PhysRevLett.108.130501. 41, 42, 57

[110] A. Peter Young, S. Knysh, and V. N. Smelanskiy, "Size dependence of the minimum excitation gap in the quantum adiabatic algorithm," *Physical Review Letters*, 101, 170503 (2008). DOI: 10.1103/PhysRevLett.101.170503. 22

Author's Biography

CATHERINE C. MCGEOCH

Catherine C. McGeoch earned her Ph.D. in computer science from Carnegie Mellon University in 1986. She joined the faculty of Amherst College in 1987, where she is presently the Beitzel Professor of Technology and Society and past chair of the computer science department. Starting in summer 2014, she has taken a leave of absence from the college to work at D-Wave Systems Inc.

Her research interests center around the development of methods and techniques for experimental analysis of algorithms, with special emphasis on algorithms and heuristics for NP-Hard problems. She co-founded the DIMACS Challenges with David S. Johnson, and co-founded the annual workshops on Algorithm Engineering and Experimentation (ALENEX), with Michael Goodrich. She is past Editor in Chief of the ACM *Journal of Experimental Algorithmics*, and currently a member of the ACM publications board. She is the author of *A Guide to Experimental Algorithmics*, published in 2012.

Printed in the United States
by Baker & Taylor Publisher Services